Accounting and Financial Management in Foodservice Operations

David K. Hayes
Jack D. Ninemeier

SENIOR EDITORIAL DIRECTOR	Justin Jeffryes
EXECUTIVE EDITOR	Todd Green
EDITORIAL ASSISTANT	Kelly Gomez
SENIOR MANAGING EDITOR	Judy Howarth
PRODUCTION EDITOR	Mahalakshmi Babu
COVER PHOTO CREDIT	© Hispanolistic/Getty Images

This book was set in 9.5/12.5 STIX Two Text by Straive™.
Published by John Wiley & Sons, Inc., Hoboken, New Jersey.
Published simultaneously in Canada.
This book is printed on acid-free paper.

Founded in 1807, John Wiley & Sons, Inc. has been a valued source of knowledge and understanding for more than 200 years, helping people around the world meet their needs and fulfill their aspirations. Our company is built on a foundation of principles that include responsibility to the communities we serve and where we live and work. In 2008, we launched a Corporate Citizenship Initiative, a global effort to address the environmental, social, economic, and ethical challenges we face in our business. Among the issues we are addressing are carbon impact, paper specifications and procurement, ethical conduct within our business and among our vendors, and community and charitable support. For more information, please visit our website: www.wiley.com/go/citizenship.

ISBN: 978-1-394-20886-9 (PBK)

Library of Congress Cataloging-in-Publication Data

LCCN is 2023041646

The inside back cover will contain printing identification and country of origin if omitted from this page. In addition, if the ISBN on the back cover differs from the ISBN on this page, the one on the back cover is correct.

SKY10057781_101823

Contents

Preface

The purpose of this book is to teach foodservice operators how they can use accounting and financial management techniques, as well as their own education, experience, values, and goals, to make the very best management decisions possible for the businesses they manage.

Every foodservice operation must offer high-quality menu items and provide excellent service to its guests, but these factors are still not enough to ensure an operation's long-term financial success. How foodservice operators account for their revenue and expenses and manage their money are critical factors to reaching their financial objectives.

It is important to understand that proper accounting alone cannot "make" the managerial decisions required of foodservice operators. It can, however, help operators make *better* decisions than those made by operators who do not understand how proper accounting procedures could help them.

The purpose of professional accounting is to report (account for) an operation's money and other valuable property. Accounting in a foodservice operation occurs even before the operation opens. Consider that operators must estimate their initial costs before they decide to open their businesses, and they often seek loans from banks or others to provide them with funds needed to open their doors.

Those who consider lending money to a foodservice operator will want to know about the estimated financial performance of the business before they lend money to it. These potential investors will also want to know that any financial estimates presented to them have been professionally considered and prepared.

Professional accounting methods are the tools used to report and analyze the financial operation of a business. Users of professionally prepared financial information include both internal (owners and operators) and external (lenders, investors, and government agencies) stakeholders. All will be interested in the financial performance of a foodservice business as reported using professional accounting and financial management practices.

This book is distinctive in that it addresses those accounting and financial management techniques that are unique to the foodservice business. Among the many key financial topics addressed in this book are:

✓ The purpose of the Uniform Systems of Accounts for Restaurants (USAR)
✓ The proper use of double-entry accounting to record business transactions
✓ How to read an income statement
✓ How to read a balance sheet
✓ How to prepare a statement of cash flows
✓ How to calculate a break-even point
✓ How to price menus for profits
✓ How to control food and beverage product costs
✓ How to manage the cost of labor
✓ How to prepare an accurate operating budget
✓ How to compare planned operating results to actual results
✓ How to develop a revenue security program
✓ How to record depreciation and account for fixed assets
✓ How to choose professional accounting assistance

Most importantly, all of these topics and more are addressed in ways specific to foodservice operations of all sizes and in all industry segments.

The ability to properly record and manage the revenue and expenses of a foodservice operation is essential to the operation's success. This is true regardless of whether the operation is a food truck, coffee kiosk, ghost restaurant, quick service restaurant, full-service operation, or a non-commercial facility.

Readers will quickly find that the content of this book is essential to the successful management of their own operations, and they will also find that the information in each chapter has been carefully selected to be easy to read, easy to understand, and easy to apply.

Book Features

In addition to the essential foodservice accounting and financial management information it contains, special features were carefully crafted to make this learning tool powerful but still easy to use. These features are:

1) **What You Will Learn** To begin each chapter, this very short conceptual bulleted list summarizes key issues readers will know and understand when they complete the chapter.
2) **Operator's Brief** This chapter-opening overview states what information will be addressed in the chapter and why it is important. This element provides readers with a broad summation of all important issues addressed in the chapter.

3) **Chapter Outline** This two-level outline feature makes it quick and easy for readers to find needed information within the body of the chapter.

4) **Key Terms** Professionals in the foodservice industry often use very special terms with very specific meanings. This feature defines important (key) terms so readers will understand and be able to speak a common language as they discuss issues with their colleagues in the foodservice industry. These key terms are also listed at the end of each chapter in the order in which they initially appeared.

5) **Find Out More** In a number of key areas, readers may want to know more detailed information about a specific topic or issue. This useful book feature gives readers specific instructions on how to conduct an Internet search to access that information and why it will be of importance to them.

6) **Technology at Work** Advancements in technology play an increasingly important role in many aspects of foodservice operations. This feature was developed to direct readers to specific technology-related Internet sites that will allow them to see how advancements in technology can assist them in reaching their operating goals.

7) **What Would You Do?** These "mini" case studies located in every chapter of the book take the information presented in the chapter and use it to create a true-to-life foodservice industry scenario. They then ask the reader to think about their own response to that scenario (i.e., *What Would You Do?*).

This element was developed to help heighten a reader's interest and to plainly demonstrate how the information presented in the book relates directly to the practical situations and challenges foodservice operators face in their daily activities.

8) **Operator's 10-Point Tactics for Success Checklist** Each chapter concludes with a checklist of tactics that can be undertaken by readers to improve their operations and/or personal knowledge. For example, in a chapter of the book related to the analysis of an income (P&L) statement, one point in that chapter's 10-Point Tactics for Success Checklist is:

> *Operator understands how to calculate earnings before interest, taxes, depreciation, and amortization (EBITDA) from a USAR-formatted income statement.*

Instructional Resources

This book has been developed to include learning resources for instructors and for students.

To Instructors

To help instructors (and corporate trainers!) effectively manage their time and enhance student-learning opportunities, the following resources are available on the instructor companion website at www.wiley.com/go/hayes/financefoodservice.

✓ Instructor's Manual that includes author commentary for "What Would You Do" mini case-study questions.
✓ PowerPoint slides for instructional use in emphasizing key concepts within each chapter.
✓ A 100-item Test Bank consisting of multiple-choice exam questions, their answers, and the location within the book from which the question was obtained. The test bank is available as a print document and as a Respondus computerized test bank. Note: **Respondus** is an easy-to-use software program for creating and managing exams that can be printed to paper or published directly to Blackboard, WebCT, Desire2Learn, eCollege, ANGEL, and other e-learning systems.

To Students

Learning about accounting and financial management in foodservice operations will be fun. That's a promise from the authors to you. It is an easy promise to make and keep because working in the foodservice industry is fun. And it is challenging. However, if you work hard and do your best, you will find that you can master all of the important information in this book.

When you do, you will have gained invaluable knowledge that will enhance your skills and help advance your own hospitality career. To help you learn the information in this book, online access to over 225 PowerPoint slides is available to you. These easy-to-read tools are excellent study aids and can help you when taking notes in class.

Acknowledgments

Accounting and Financial Management in Foodservice Operations has been designed to be the most up-to-date, comprehensive, technically accurate, and reader-friendly learning tool available to those who want to know how to increase profits by effectively managing the financial components of their foodservice operations.

The authors thank Catriona King of Wiley for initially working with us to develop the idea for a series of practical books that would help foodservice operators of all sizes more effectively manage their businesses. She was essential in helping conceptualize the need for this book as well as all of the other books in this five-book *Foodservice Operations* series. The five titles in the series are:

✓ *Accounting and Financial Management in Foodservice Operations*
✓ *Marketing in Foodservice Operations*
✓ *Cost Control in Foodservice Operations*
✓ *Managing Employees in Foodservice Operations*
✓ *Successful Management in Foodservice Operations*

We would also like to thank the external reviewers who gave so freely of their time as they provided critical industry and academic input on this series. To our reviewers, Dr. Lea Dopson, Gene Monteagudo, Isabelle Elias, and Peggy Richards Hayes, we are most grateful for your comments, guidance, and insight. Also, thanks to Michael T. Kavanagh, who was a technological friend indeed, when we were most in need!

Books such as this require the efforts of many talented specialists in the publishing field. The authors were extremely fortunate to have Todd Green, Judy Howarth, and Kelly Gomez at Wiley as our publication team. Their efforts went far in helping the authors present the book's material in its best and clearest possible form.

Finally, the authors would like to thank the many students and industry professionals with whom we have interacted over the years. We sincerely hope this book allows us to give back to them as much as they have given to us.

David K. Hayes, Ph.D.
Jack D. Ninemeier, Ph.D.

Dedication

The authors are delighted to have the opportunity to dedicate this book, and this entire *Foodservice Operations* series, to two outstanding and unique individuals.

Brother Herman Zaccarelli

Brother Herman E. Zaccarelli, C.S.C., passed away in 2022 at the Holy Cross House in Notre Dame, Indiana. His professional work included many projects for the hospitality industry, and he published several books and hundreds of articles for numerous trade publications over many years. Among numerous accomplishments, Herman founded Purdue University's Restaurant, Hotel, and Institutional Management Institute in 1976. Later, he served as Director of Business and Entrepreneurial Management at St. Mary's University in Winona, Minnesota.

A lifelong learner, at the age of 68, Brother Herman retired to Florida where he earned a bachelor's degree in Educational Administration and a master's degree in Institutional Management.

Herman's ideas and concepts have been widely adopted in the hospitality industry, and he assisted many young educators including the authors of this book series. He will be remembered as a colleague with creative ideas who provided significant assistance to those studying and managing in the hospitality industry. Herman was especially helpful in discovering and addressing learning opportunities for Spanish-speaking students, educators, and managers throughout the United States and around the world.

Dr. Lea R. Dopson

A lifelong friend, advisor, and colleague, as well as an outstanding author herself, at the time of her untimely passing, Lea served as President of the International Council on Hotel, Restaurant, and Institutional Education (ICHRIE) and Dean of the prestigious Collins College of Hospitality Management (Cal Poly Pomona).

Lea was a dedicated hospitality professional and a fierce advocate for hospitality students at all levels. Those who knew her were continually in awe of her intelligence and humility.

It was especially fitting that Lea was named as a recipient of the H.B. Meek Award. That award is named after the individual who started the very first hospitality program in the United States (at Cornell University). Selected by the recipient's peers, it goes *not* to the most outstanding academic professional working in the United States but to the most outstanding academic professional in the entire world. That was Lea.

While she is dearly missed, her inspiration goes on everlastingly in the works of the authors.

1

Fundamentals of Accounting and Financial Management

What You Will Learn

1) The Purpose of Professional Accounting
2) The Language of Professional Accounting
3) The Mechanics of Financial Management

Operator's Brief

In this chapter, you will begin to learn about the tools and procedures used to effectively manage a foodservice operation's finances. Professional accounting methods are the tools used to report and analyze the financial operation of a business. Users of professionally prepared financial information include both internal (business owners and operators) and external (lenders, investors, and government agencies) stakeholders. All will be interested in the financial performance of a foodservice business as reported using professional accounting practices.

There are several accounting specialties including financial accounting, auditing, tax accounting, cost accounting, and managerial accounting. The focus of this book is on managerial accounting: the process of using historical and estimated financial information to help those who manage foodservice operations better plan for the future of their businesses.

In addition to the operators of a foodservice business, there are likely others who require professionally prepared accounting information. These include owners, boards of directors, creditors, government agencies, employee unions, and financial analysts. Since so many entities are interested in the financial performance of a foodservice operation, it is recommended that the Uniform System of Accounts for Restaurants (USAR) be used to prepare financial information.

(Continued)

To be most easily understood, those responsible for business accounting in any industry use a consistent language to report financial results. These are called Generally Accepted Accounting Principles (GAAP), and they are utilized when providing information about a foodservice operation's revenues, expenses, and profits.

The mechanics of a professional accounting and financial management system include the three key components of bookkeeping, financial summary production, and financial analysis. Since transparency is important, those responsible for accounting in a foodservice operation must also recognize the ethical considerations of presenting financial information about a business.

CHAPTER OUTLINE

Professional Accounting
    Accounting for Effective Financial Management
    Accounting Specializations
    Users of Accounting Information
    The Uniform System of Accounts for Restaurants (USAR)
Generally Accepted Accounting Principles (GAAP)
The Mechanics of Financial Management
    Bookkeeping
    Summary Accounting
    Financial Analysis
Ethics in Accounting

Professional Accounting

Some foodservice operators feel that **accounting** for their businesses is an extremely complex process. While professional accounting does require attention to detail, every foodservice operator can master the basic skills required to effectively record and analyze their financial results.

The term "accounting" actually originated from an old Middle French word *acompter*, which itself originated from Latin *ad + compter*, meaning "to count." Since all foodservice operators know how to count, they can master the most important accounting principles even if they do not consider themselves to be a professional **accountant**!

Key Term

Accounting: The system of recording and summarizing financial transactions and analyzing, verifying, and reporting the results.

Key Term

Accountant: An individual skilled in the recording and reporting of financial transactions.

Accounting for Effective Financial Management

The purpose of professional accounting is to report (account for) an operation's money and other valuable property. Professional accounting is utilized by all managers in business including those in the foodservice industry, and accounting principles are utilized every time a guest purchases food or beverages.

Accounting in a foodservice operation occurs even before the operation opens. Consider that operators estimate their initial costs before they decide to open their businesses, and they often seek loans from banks or others to provide them with funds needed to open their doors.

Those who consider lending money to a foodservice operator will want to know about the estimated financial performance of the business before they lend money to it. These potential investors will also want to know that any financial estimates presented to them have been professionally considered and prepared.

Professional and accurate accounting is important to many individuals in the foodservice industry. The owners of an operation will want to monitor their business's financial condition. These owners may be one or more individuals, partnerships, or small or very large corporations, but they all care about the performance of their investments.

Investors in a foodservice operation generally want to put their money in businesses that will conserve or increase their wealth. To monitor whether their investments are good ones, investors seek out and rely upon accurate financial information. When it is done properly, the professional accounting process provides that important information.

Professional accounting is actually a large field of study. To understand why accounting plays such a significant role in business, consider just a few examples of the type of basic and important questions the discipline of accounting can readily answer for foodservice managers:

1) What was the total sales level achieved by our business last month?
2) How many guests did we serve?
3) What was our most popular menu item?
4) What percentage of our revenue was achieved from sales of alcoholic beverages?
5) What **guest check average** was achieved last week? Was it higher or lower than the prior week?
6) What portion of our revenue is being spent on labor?
7) What percentage of our sales was achieved through our take-out and delivery services?

Key Term

Guest check average: The average (mean) amount of money spent per guest (or table) during a specific accounting period. Also referred to as "check average" or "ticket average."

8) Are our operating expenses higher or lower than those of other operations of a similar type?
9) Are we more or less profitable this month than last month?
10) What is our operation realistically worth if we were to sell it today?

The above are just a sample of the many questions hospitality managers utilize to answer accounting-related issues. However, it is important to understand that accounting is not the same as management. Rather, professional accounting is a tool used by effective foodservice operators to manage their businesses.

To better understand the difference between professional accounting and professional management, consider these examples of questions that are *not* best answered by using accounting information alone:

1) Should I promote Jenny or Raul to train our new dining room wait staff?
2) Should the portion size of the beef patty used to make our signature "Patty Melt" be 5 ounces or 7 ounces?
3) Would our guests prefer an increase in the number of inexpensive or higher-quality (but more costly) wines when we create our new wine list?
4) We currently close at 10:00 p.m. on Friday and Saturday night. Would we attract more customers if we decided to stay open an extra hour on those nights?
5) What are the best packaging materials to use for our carryout and delivery orders?
6) Should we assign someone additional hours to make more regular postings on our Facebook account? Who on our current staff is best qualified to perform this task?

Note that, in each of these questions, the best decision requires foodservice operators to utilize their own experience and best judgment of what is "right" for their guests, their employees, and their businesses. As a result, accounting alone cannot make the decisions called for in the questions above. It can however, when properly used, help operators make *better* decisions about these types of issues than those made by operators who do not understand how accounting could help them.

The purpose of this book is to teach operators how they can use accounting techniques and their own education, experience, values, and goals to make the very best management decisions for the businesses they manage.

Foodservice managers use properly prepared financial information to manage activities involving money that is earned and spent in the business. Financial information that summarizes these activities must be organized and expressed in meaningful ways. Analysis and interpretation of this data is necessary, and the results must be recorded, summarized, and reported to those needing to know about the economic health of the operation.

Users of professionally prepared financial information will be both internal (the business owners and managers) and external (lenders, investors, and government agencies). Financial management and professional accounting are not the same as **bookkeeping**, and there is a big difference between the two.

Key Term

Bookkeeping: The process of recording a foodservice operation's financial transactions into organized accounts on a daily basis.

Financial management includes organizing, analyzing, interpreting, recording, summarizing, and reporting financial information. By contrast, a bookkeeper's primary task is to analyze and *record* financial transactions. Note: In large foodservice organizations, a bookkeeper may handle only one type of transaction such as sales, accounts payable, or payroll. The organization's accountant would then summarize the bookkeeper's work and further interpret the results for the organization's owners and managers.

Accounting Specializations

As shown in Figure 1.1, there are several specialized areas within the accounting profession. In this section, we will explore details for each of the specialized areas.

Figure 1.1 Accounting Specializations

Financial Accounting

Financial accounting involves the overall process of developing and using accounting information to make good business decisions.

Financial accountants prepare financial statements including the income statement, balance sheet, and statement of cash flows all of which (and more!) are addressed in this book. These are among the most important reports that owners, managers, government agencies, financial institutions, and others use to learn about the current financial status of a restaurant.

To better understand how financial accounting can help foodservice operators, consider Tonya Richards. She is interested in starting her own small pizza shop. The shop would be located in a strip shopping center and would sell primarily pizzas, hot subs, and soft drinks. Some of the many financial considerations confronting Tonya as she tries to determine whether buying and operating the shop is a good idea include:

1) How much revenue do pizza shops like this typically achieve on an average day? An average month?
2) What do pizza shops normally spend to properly staff their stores?
3) How much should I spend on the equipment needed to prepare the menu items I will sell?
4) Given the size and location of my store, what is a reasonable price to pay for obtaining insurance for my business?
5) How much money am I likely to make for myself during the first year I own the store? How much in future years?

Tonya can get important information from her financial accountant, but she will also need managerial skills and her own intuition and talents to provide answers to these and other business questions she must answer.

Auditing

Auditing is an independent verification of financial records, and an **auditor** is an individual or group of individuals that completes the verification. The accurate reporting of financial transactions is important to many different entities including managers, owners, investors, and taxing authorities. The auditing branch of accounting is chiefly concerned with the accuracy and truthfulness of financial reports. It is also concerned with safeguarding the assets of a business from those unscrupulous individuals who would defraud or otherwise take advantage of it.

Properly performed, the auditing branch of accounting is designed to point out accounting weaknesses and irregularities and help to prevent accounting fraud.

In part because of the potential damage that could be done by unscrupulous business operators, in 2002 the U.S. Congress passed the Sarbanes–Oxley Act (SOX). Technically known as the Public Company Accounting Reform and Investor Protection Act, the law provides criminal penalties for those who have committed accounting fraud. Sarbanes–Oxley covers a whole range of corporate governance issues including the regulation of auditors assigned the task of verifying a company's financial health. Ultimately, Congress determined that a company's implementation of proper accounting techniques was not merely good business. Instead, it would be the law, and violators would be subject to fines or even prison terms.

Not surprisingly, SOX has revised the role of auditors responsible for conducting an audit of a foodservice operation's accounting methods and techniques, and their role has become increasingly important. Individuals who are directly employed by a company to examine the company's own accounting procedures are called internal auditors, and they can play a valuable role in assessment because they usually understand the company's business so well. Note: External auditors (individuals or firms hired specifically to give an independent assessment of a company's compliance with standardized accounting practices) can also be retained.

Many foodservice operators serve as their own in-house auditors. If the facility they manage is part of a larger company or chain of units, their company may also employ in-house auditors. Auditors help to ensure honesty in financial reporting as they devise the systems and procedures needed to help protect and safeguard business assets. As a result, hospitality managers use auditors and auditing techniques to help address many internal questions including:

1) Are all our purchases supported by the presence of a legitimate invoice before we process payment?
2) Are any guest adjustments from their initial bill supported by written documentation explaining why the bill was adjusted?
3) Is all the revenue reported by our business fully documented and compared and matched to bank statements that list actual deposits made into the business's bank accounts?

4) Are wages paid to employees supported by a written and verifiable record of hours worked?

5) Do we have accurate records showing compliance with all requirements imposed by local, state, and federal taxing authorities?

The best auditors help ensure that financial records are accurate. They also assist operators in reducing waste and preventing fraud.

Find Out More

The 2002 SOX became law to help rebuild public confidence in how corporate America governs its business activities. The act has far-reaching implications for the tourism, hospitality, and leisure industries. To examine an overview of its provisions, enter www.sec.gov/about/laws.shtml in your favorite browser. When you arrive at the site choose "Sarbanes–Oxley Act of 2002."

Tax Accounting

Tax accounting is the branch of accounting that concerns itself with the proper and timely filing of tax payments, forms, and other required documents with the governmental units that assess taxes. Professional tax accounting techniques and practices ensure that businesses properly fulfill their legitimate tax obligations.

In the foodservice industry, all operators are required to implement systems that will carefully record any taxes that will be owed by their businesses.

Key Term

Tax accounting: The accounting specialty that involves planning and preparing for required tax payments and filing tax-related information with governmental agencies.

Key Term

Cost accounting: The accounting specialty that is involved with classifying, recording, and reporting a business's expenses or costs.

Cost Accounting

Cost accounting is the branch of accounting concerned with classifying, recording, and reporting a foodservice operator's expenses or costs. All businesses seek to control their costs and not waste their money, and those who manage foodservice operations are very concerned about their operating costs.

Cost accountants can determine costs by different departments such as the kitchen and the bar, by delivery service style such as dine-in or take-out, and by the guest services or merchandise sold by a business. They create systems to classify costs and report them in ways that are most useful to those who need to know how a business spends its money.

Find Out More

The control of costs is important to every business and especially so in the foodservice business. Professionals working in the foodservice industry have a variety of tools available to them to help manage their costs. One of the very best is the book *Food and Beverage Cost Control,* written by Dr. Lea Dopson and Dr. David Hayes, and published by John Wiley.

To review the contents of this extremely popular book, go to www.wiley .com. When you arrive at the Wiley website, enter *Food and Beverage Cost Control* in the search bar to examine the outline and content of this book's most recent edition.

Managerial Accounting

Managerial accounting is the specialty area of accounting that is the primary topic of this book. Managerial accounting is the specialization that helps managers make decisions about the future. To better understand the purpose of managerial accounting, assume that Mariana Gomez is the person responsible for providing in-flight meals to international travelers on flights from New York City to Paris. She manages a large commercial kitchen located near the John F. Kennedy Airport.

Key Term

Managerial accounting: The accounting specialty that uses historical and estimated financial information to help foodservice operators plan the future.

Mariana's clients are the airlines who count on her operation to provide passengers with tasty and nutritious meals at a per-meal price the airlines find affordable.

One of Mariana clients wishes to add a new daily flight beginning next month. The evening flights will carry an average of 500 travelers, each of whom will be offered one of two in-flight meal choices for dinner. The client would like to provide each flier with a choice of a beef or a chicken entrée. To ensure that the maximum number of fliers can receive their first choice, should Mariana's operation plan to provide each flight with 500 beef and 500 chicken entrées? (The answer, most certainly, is No!)

To prepare 1,000 meals (500 of each type) would indeed ensure that each traveler would always receive his or her first meal choice, but it would also result in the production of 500 wasted meals (the 500 meals *not* selected) on each flight. It would be difficult for Mariana to provide the airline with cost-effective per-meal pricing when that many meals are inevitably wasted.

The more cost-effective approach would be to accurately forecast the number of beef and chicken entrées that would likely be selected by each group of passengers and then produce that number. The problem, of course, is in knowing the optimum number of each meal type that should be produced. If Mariana had carefully and consistently recorded previous meal-related transactions (entrées chosen by fliers on previous flights), she would be in a much better position to use this information. She could, for example, estimate the actual number of each entrée the new passengers would likely select. If she had done so, she would be using managerial accounting.

Managerial accounting is the system of recording and analyzing transactions for the purpose of making management decisions of this kind. It consists of utilizing accounting information (historical records in this specific case) to make informed management decisions.

Managerial accounting is one of the most exciting accounting specializations. Its proper use requires skill, insight, experience, and intuition, and these are the same characteristics possessed by the best foodservice operators. As a result, excellent foodservice operators most often become excellent managerial accountants.

There are a number of different accounting specializations, and each is very important. These specializations are summarized in Figure 1.2. Foodservice operators may find that they perform some tasks in each of these specializations, or they may choose to employ individuals or organizations to help them. In all cases, each of the specializations in accounting can provide important information.

Users of Accounting Information

Foodservice operators are, of course, interested in the financial performance of their businesses and use the various specializations of accounting to assess that performance. However, foodservice operators are not the only ones interested in the financial performance of their businesses. There are other important users of accounting information.

Owners and Investors

Owners are those who have invested in the operation and may include one person in a sole proprietorship, two or more people in a partnership, or up to thousands

Specialization	Purpose
1) Financial	Record financial transactions
2) Auditing	Verify accounting data and procedures
3) Tax	Compute taxes due
4) Cost	Identify and control costs
5) Managerial	Make management decisions using accounting information

Figure 1.2 Specializations of Professional Accounting

of people in a corporation. All of these owners and investors will want to know how their investments are doing.

Boards of Directors
Large foodservice operations or multi-unit chains may have corporate stockholders who elect persons to represent them in the management of the business. These individuals need accurate accounting information to evaluate the effectiveness of the managers in charge of foodservice operations.

Creditors
Those who lend money (banks and other lenders) or who provide products and services (vendors and suppliers) will want to know the likelihood that the payment obligations of a foodservice operation will be met in full and on time before they agree to extend credit to the operation.

Government Agencies
Income earned by a foodservice operation is taxable by the federal government, most states, and many communities. For example, the Internal Revenue Service (IRS) at the federal level, state revenue departments, and local taxing authorities have an on-going interest in a foodservice operation's accounting records. Also, the Securities and Exchange Commission (SEC), a federal agency, must review audited financial statements as it approves prospective information developed by large restaurant organizations before they issue company stock to the general public.

Employee Unions
Some foodservice operations are unionized. Accounting information is used by union officials and members in unionized foodservice operations to assess the abilities of a business to meet wage and benefit requirements.

Financial Analysts
Persons outside of a foodservice operation such as staff members of mutual investment and insurance companies may desire accounting information about an operation for their own or a client's purposes.

The Uniform System of Accounts for Restaurants (USAR)

As noted above, many individuals and groups are interested in a foodservice operation's financial performance. Therefore, this information must be prepared in a manner that is consistent, easily read, and easily understood.

Laws exist requiring owners to properly report and pay taxes due to file certain documents with the government and to supply accurate business data to various other entities. As a result, many hospitality companies require that their managers use a series of standardized (uniform) accounting procedures. These are called a **uniform system of accounts** and simply represent agreed-upon methods of recording financial transactions.

Business owners use a uniform system of accounts to provide uniformity and consistency in reporting financial information about their businesses, and a uniform system of accounts provides accuracy, reliability, and comparability. The financial information produced using a uniform system of accounts will be verifiable and auditable by a third-party, and it will be understood by lenders, investors, taxing authorities, and others.

Different businesses have different accounting needs, and there are uniform systems of accounts produced specifically for individual business segments. In the hospitality industry, some of the best known of these uniform systems are the:

✓ **Uniform System of Accounts for Restaurants (USAR)**
✓ Uniform System of Financial Reporting for Clubs (USFRC)
✓ Uniform System of Accounts for the Lodging Industry (USALI)

Uniform accounting systems are continually reviewed and periodically revised. For example, this book was prepared using reporting principles contained in the eighth and most current edition of the USAR. Important specific recommendations of the USAR will be addressed in detail in the appropriate portions of this book.

To illustrate why the use of the USAR is so important, assume that an individual owned two Italian restaurants located in two different cities, and both offered the same menu. The owner wants to assess the ability of the two individuals responsible for operating the restaurants. It would be very confusing if the units' two managers used different methods for preparing and reporting each of their operations' financial results. In this example, unless managers both report and account for their financial performance in a way that is consistent (uniform), the performance of the two operations and their managers could not be properly analyzed and compared to each other.

The use of the USAR to produce accounting information for foodservice operations is not mandatory, but use of the applicable USAR is highly recommended. One reason is that a primary purpose of preparing accounting information is to

Key Term

Uniform system of accounts:
Accounting standards used to provide uniformity and consistency in reporting financial information.

Key Term

Uniform System of Accounts for Restaurants (USAR): A recommended and standardized (uniform) set of accounting procedures used for categorizing and reporting restaurant revenue and expenses.

clearly identify revenue, expenses, and profits for a specific time period. The best foodservice operators want to do this properly so the financial records of their businesses will accurately reflect their efforts.

What Would You Do? 1.1

"I know what Robert asked me to do, I'm just not sure how to do it!" said Sally.

Sally, the assistant manager at Chez Paul's French restaurant was talking to Otis, the restaurant's dining room manager.

"What's the problem?" asked Otis.

"The problem," replied Sally, "is that our manager Robert asked me to do the weekly inventory while he's on vacation. I'm supposed to count the items we have in inventory, multiply the amount we have on hand by the cost of the items, and determine our total inventory value."

"That seems easy enough," said Otis.

"Well the counting part is easy," said Sally, "and I know how to multiply. What I don't know is how much to multiply by. For example, we have 180 pounds of strip steaks in inventory. Some of them were purchased at $14 a pound and some at $18 a pound."

"So maybe multiply by $16 a pound?" questioned Otis.

Assume you were the owner of Chez Paul's. How important do you think it would be for inventory valuations produced by Sally to be calculated in the same way as those previously taken by Robert? What would be the likely result if they were not?

Generally Accepted Accounting Principles (GAAP)

The way operating statistics are compiled, and revenue or expense data is reported can be very significant to correctly interpret that data. Professional accountants in the foodservice industry use very specific principles to prepare financial information related to a business's revenue, expense, and profits and communicate results to others.

Those who read financial information must have faith that the collection and reporting systems used to prepare it are accurate and consistent. The information presented must also be easily understood. **Generally Accepted Accounting Principles (GAAP)** constitute the framework against which proper accounting procedures and techniques are measured.

Key Term

Generally Accepted Accounting Principles (GAAP): Standards that have evolved in the accounting profession to ensure uniformity in the procedures and techniques used to prepare financial statements.

Some of the most important GAAP in use and that directly affect foodservice operators include:

1) **Business entity:** This GAAP states that a foodservice operation is a distinct business separate from its owners. It generates revenue, incurs expense by using **assets**, and makes a profit, suffers a loss, or "breaks-even" by, and for, itself. This distinct business principle is important because it states a business's financial records cannot be combined with the personal financial records of the business's owners. This is true even if a single individual owns the business.

   The impact of this distinct principle occurs when income is measured as it is generated by the business, not when it is distributed to owners. Likewise, an obligation owed by the business is considered a **liability**. A liability may be money owed to a vendor to pay for products already received or as noted above, the liability may be an obligation such as loan to the business that the owners owe to themselves!

2) **Historical cost:** The value of a business asset is its agreed-upon cash equivalent. When a transaction occurs, for example, when an asset such as a piece of kitchen equipment is purchased, the price paid for it should reflect its **current fair value**.

   Over time, the value of an asset may change. For example, inflation may increase the value of land or buildings. However, the **historical cost**, not the current fair value, normally represents the asset's value in an operation's accounts and in its financial statements.

   To illustrate the historical cost principle, assume a building costs a foodservice operator $6,000,000 to buy. At the time of purchase, the cost would be equal to the building's current fair value. Three years later, however, the current fair value of the building may have gone up to $7,000,000, but the operator's financial statements would still reflect the historic cost of $6,000,000.

Key Term

Assets: Something of value owned by a foodservice operation. Examples include cash, product inventories, equipment, land, and building(s).

Key Term

Liability: Obligations (money owed) to outside entities. Examples include amounts owed to suppliers for delivered products, to lenders for long-term debt such as a mortgage, and to employees (payroll) that has been earned by, but not yet paid to, an operation's staff members.

Key Term

Current fair value: The measuring of a business's liabilities and assets at their current market value; the amount that an asset could be sold for (or that a liability could be settled for) that is fair to both buyer and seller.

Key Term

Historical cost: A measure of value used in accounting in which the value of an asset is recorded at its original cost when acquired by the company. The historical cost method is used for fixed assets under Generally Accepted Accounting Principles (GAAP).

3) **Going concern:** Accountants assume, unless there is reason to believe otherwise, that an operation will exist in the indefinite future. If, for example, the operation were to cease doing business, certain liabilities would be due immediately. Likewise, assets might need to be sold at a considerable loss. When accountants assume that a business will continue (and this is the normal assumption), there is no need to write down assets to a liquidation value or to reclassify long-term liabilities as being due immediately.

4) **Periodicity:** Periodicity, or the "time period" principle is important because it requires a business to clearly identify the dates for which its financial transactions are reported. For the owners of a business, it is likely that the most important financial reports they will examine are those that include all of the financial transactions occurring during their **fiscal year.**

> **Key Term**
>
> **Fiscal year:** A time period that can begin on any date and then concludes 365 consecutive days after it begins.

A fiscal year, which consists of 12 consecutive months (but not necessarily beginning in January and ending in December like a **calendar year**), would most likely include the business's best, as well as its poorest, periods of financial performance. A fiscal year-end financial report will provide owners with the information needed to file their taxes and perhaps make other needed financial management decisions.

> **Key Term**
>
> **Calendar year:** A 365-day time period that begins on January 1st and ends on December 31st of the same year.

The amount of time included in any summary of financial information is called an **accounting period**. The managers of a business may be most interested in monthly, weekly, or even daily financial summary reports. These would inform the managers of the business's revenue and expense levels

> **Key Term**
>
> **Accounting period:** The amount of time included on a financial summary or report, and which should be clearly identified on the financial document.

and profitability. The owners of the business would also be interested in such reports. Regardless of the report's readers, however, it is critical that the financial statements clearly state the time period they address.

In some cases, the time period reported in a financial statement can be somewhat arbitrary. For example, consider foodservice operations that are open 24 hours a day, 7 days a week (the businesses never close). However, their managers must still select a point in time (usually between midnight and 5:00 a.m. when the operation will likely have the fewest financial transactions) to end the recording of one business day and begin recording the next day's financial transactions. In a properly prepared financial summary of any business, the periodicity principle assures that the reader will be clearly informed of the precise time period included in the summary.

5) **Expenses matched to revenue:** This GAAP requires that the **expenses** of a foodservice business be matched with and deducted from the **revenues** that are generated using an **accrual accounting system.** This accounting system recognizes revenues and expenses without concern for when cash is received or paid out by the business.

In an accrual accounting system, money owed to a foodservice operation is called **accounts receivable (AR)**, and money owed by the operation to suppliers and others is referred to as **accounts payable (AP)**.

Some very small foodservice operations may use a cash accounting system, which treats revenues as income when cash is received, and expenditures as expenses when cash is paid out. However, GAAP requires the use of an accrual accounting system, and this system will be the basis for the accounting discussions throughout this book.

6) **Consistency:** The consistency principle states that the same procedures used to collect and report accounting information are used each fiscal period. If this GAAP were not used, those who read about an operation's financial performance would not have accurate and uniform information upon which to base their decisions.

7) **Materiality and practicality:** The matching and consistency principles previously addressed require accountants using an accrual system to do their best to match a business's expenses with the time period in which those expenses were incurred. The materiality principle, however, allows accountants, under very strict circumstances, to vary from these two important principles.

Key Term

Expense: A decrease in a resource, such as food inventory, which occurs when a foodservice operation sells a product or service or incurs a business cost.

Key Term

Revenue: An increase in a resource such as cash, which occurs when a product or service is sold by a business. Also commonly referred to as "sales" or "income."

Key Term

Accrual accounting system: An accounting system that matches expenses incurred with revenues generated. This is done with the use of accounts receivable, accounts payable, and other similar accounts.

Key Term

Accounts receivable (AR): Money owed to a foodservice operation, generally from guests, which has not yet been received. Sometimes referred to as "AR."

Key Term

Accounts payable (AP): Money owed by the foodservice operation to suppliers and lenders that has not yet been paid. Sometimes referred to as "AP."

To better understand the reason for this GAAP, consider the foodservice operator who buys a $15 stapler for use in the operator's office. Assume that the stapler has an expected useful life of five years. In this situation, it could be argued that the operation's accountants should charge the restaurant's administrative expense account $0.25 each month for each of the next 60 months ($15.00 expense/60 months stapler life = $0.25 per month) to properly account for the stapler's expense.

The principle of materiality, however, states that in a case such as this, the amount of money involved is so small, the operation's accountants can expense the entire cost of the item in the same month it was purchased. To do otherwise would cause unnecessary work and would be unlikely to materially (significantly) affect the outcome of the operator's monthly expense statement.

The materiality principle, then, means that, if an item is deemed to not be significant, other accounting principles may sometimes be ignored when it is not practical to apply them. Stated another way, since businesses seek to make a profit, *it makes little sense to spend a lot of dollars accounting for a few pennies!*

The amount of money that must be involved before an expense is considered "material" can vary for each business and should be clearly identified. The amount of money expended before it is material should be unmistakably established by a business's owners, carefully applied to the business's records by its accountants, and freely shared with readers of the business's financial statements.

8) **Conservatism:** This GAAP requires that all business losses should be shown in financial records if there is a reasonable chance a problem will occur. Gains and related financial benefits, however, should not be reflected in financial records until they actually happen.

To illustrate, assume a foodservice operation had a lawsuit for negligence filed against it, and the operation's legal advisor indicates the operation is likely to lose the lawsuit and can reasonably estimate the amount of the loss. The conservatism principle dictates the recording of the loss rather than waiting for the judge's decision.

This principle is important since many accounting decisions do not have a single "right" or "wrong" answer. This principle guides accountants confronted with alternative measurements to select the option that will yield the *least* favorable impact upon the business's profitability and financial position within a specific accounting period.

9) **Objectivity:** While all GAAP have value, the objectivity principle most ensures readers of a business's financial statements that the documents will be reliable. Essentially, the objectivity principle states that recorded financial transactions must have a confirmable (objective) basis in fact. That is, there must be a way to verify that a financial transaction actually occurred before it can be recorded in the business's financial records.

To illustrate, assume a foodservice manager reported that a specific day's total sales equal $5,000. The objectivity principle states that these sales must have substantiating evidence to prove that they actually occurred. In this example, that evidence could include the individual guest checks used by wait staff to record guest orders, bank card statements showing charges made to guest's cards, or various sales records maintained in restaurant's **point-of-sale (POS) system**. In each of these cases, there is objective and verifiable records that confirm the restaurant manager's assertion that the day's sales equaled $5,000.

In a similar manner, expenses must be verifiable before they can be recorded as having been incurred or paid. Examples of methods used to verify the recording and payment of invoices can include delivery slips or original invoices supplied by vendors, canceled checks, or documented electronic fund transfers (EFTs) showing that funds have been paid or moved electronically from the business to the entity to whom money is owed.

> **Key Term**
>
> **Point-of-sale (POS) system:** An electronic system that records foodservice customer purchases and payments, as well as other operational data.

Additional types of verifiable expense evidence may include written contracts showing payments are due from the business on a regular basis. Examples of this type of verification include contracts spelling out the terms and length of lease payments or loan re-payments.

Recall that auditors are accountants who ensure that standards of financial reporting are maintained in a business. As a result, one of an auditor's main tasks is ensuring that businesses report only financial data that has a basis in fact. If businesses were allowed to estimate, rather than directly confirm, revenue and expenses, the financial summaries produced by these businesses would also be estimates. It follows, then, that these summaries would not precisely reflect the money earned and expenses incurred by the business. Not surprisingly, when auditors uncover significant accounting "scandals," it is usually a violation of this principle (such as a business reporting non-existent revenues or inappropriately documenting its expenses) that generates the scandal.

10) **Full disclosure:** Bookkeeping and accounting most often report financial events that have happened in the past and simply must be recorded. For example, a lunch sale made by a Greek deli on Monday will be reported by the operation's management on Monday night. In a similar manner, a hospital

foodservice director who buys new hot food plate covers on the tenth of the month will likely summarize and report that purchase at the end of the month. Also, a food truck operator who replaces the truck's deep-fat fryer will, using appropriate accounting principles, record this expense over the useful life of the fryer (a time period that will likely span several years).

In each of these cases, when accountants rigorously apply GAAP, readers of the financial statements prepared for these businesses can have confidence that all transactions described have, in fact, been reported properly. The full disclosure principle, however, requires that accountants do even more.

This important principle requires that any past or even *future* event that could materially affect the financial standing of the business and that cannot be easily discerned from reading the business's financial statements must be separately reported. These reports, prepared in the form of footnotes, must be attached to the financial statements prepared by the business's accountants.

Accountants use full disclosure footnotes to report events that have not yet happened but could happen. And, if they do occur, they could considerably change the conclusions drawn by readers of a business's financial statements. Significant lawsuits filed against a business are just one example of a future event that must be revealed under the full disclosure principle. Other events affecting how financial statements are interpreted include changing from a cash to an accrual accounting system and significant tax disputes that occurred after the financial statements were prepared that could materially affect the business in the future.

Technology at Work

While very large foodservice operations will likely have a full-time accountant on staff or hire an accounting firm to assist them in producing their financial records, owners of smaller operations may elect to do all or most of their accounting themselves.

Fortunately, several companies have created accounting software programs specifically designed for foodservice operators. Using accounting software can help owners and managers know where their money is going and identify potential savings. The software can also track inventory and profits and properly calculate sales taxes to avoid fines.

To examine some accounting packages available to foodservice operators and learn about the costs associated with using them, enter "accounting software for restaurants" in your favorite search engine and review the results.

The Mechanics of Financial Management

As shown in Figure 1.3, the financial management of a foodservice operation consists of three key components. It is essential that all foodservice operators understand the purpose and importance of each component.

Bookkeeping

Earlier in this chapter, bookkeeping was defined as the process of recording a foodservice operation's financial **transactions** into organized accounts on a daily basis.

Key Term

Transaction: Any business event having a monetary impact on the financial statements of a business.

Proper bookkeeping forms the foundation of accurate financial reporting and analysis. It is not possible to make a meaningful analysis of a foodservice operation's financial standing if the data to be summarized was erroneously or carelessly supplied by those performing bookkeeping tasks. In addition, foodservice managers will not be able to properly analyze financial information and make correct decisions if the accounting information summarized was inaccurate.

To better understand the financial management process, consider that the recording of an individual financial transaction such as the sale of a cup of coffee is actually a bookkeeping task. It is typically completed by a server using an operation's POS system. An operator may then record the total number of cups of coffee sold in a specified time (accounting) period based on POS information. These sales as well as the other sales achieved by the operation will be summarized in

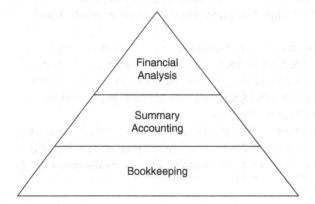

Figure 1.3 Mechanics of Financial Management

monthly financial statements to be analyzed by the operation's accountants, managers, and owners.

In the foodservice industry, the actual distinctions between bookkeeping and summary accounting are not always clear cut. Many foodservice operations are small, and bookkeeping, summary accounting, and financial analysis (managerial accounting) may all be done by only one or two individuals. For this reason, this book will not make a significant distinction between bookkeeping and accounting. For foodservice operators, however, it is important to ensure that precise and timely bookkeeping (recording) methods are used to produce the accurate financial data needed to make good decisions.

Summary Accounting

Foodservice operators should regularly create financial summaries of the transactions that have occurred in their businesses. The specific summaries created will vary based on the needs and desires of the business's owner. The following three major summary financial statements, however, are required under GAAP:

✓ Income statement
✓ Balance sheet
✓ Statement of cash flows (SCF)

The production of each of these three required accounting summaries (and more!) will be addressed in detail in the chapters that follow.

Financial Analysis

The regular and proper analysis of a foodservice operation's financial summaries are key to the operation's success. Foodservice operators frequently use a five-step process to analyze the financial performance of their businesses.

Step 1: Performance standards (expectations) are established. This is typically done through the development of an operating budget (see Chapter 8).
Step 2: Actual financial information is collected and summarized to measure operating results.
Step 3: Comparisons are made between expected performance (Step 1) and actual performance (Step 2).
Step 4: Corrective action must be taken when necessary to identify causes and bring actual results (Step 2) in line with the expected performance (Step 1).
Step 5: An evaluation of the results of corrective actions that have been taken (Step 4) is made.

Technology at Work

Due to the nature of their business, foodservice operations typically generate hundreds or even thousands of transactions a day. Recording these transactions and summarizing them creates a tremendous amount of financial data including guest payment information.

In some cases, a foodservice operator will elect to store all of that data on-site. Increasingly, however, the sheer volume of data generated requires extraordinary amounts of data storage capabilities. As a result, increasing numbers of foodservice operators are turning to the cloud for storing their financial information. Doing so has a number of advantages including 24-hour availability, heightened data security, accessibility from multiple locations, and low cost.

To examine some of the advantages of utilizing cloud-based storage for a foodservice operation's financial data, enter "cloud storage for accounting data" in your favorite search engine and review the results.

Ethics in Accounting

The foodservice industry is one of the most exciting and rewarding industries in the world. It will continue to offer its members solid employment opportunities and serve as the backbone of many local economies. However, the industry's reputation is partially built upon the knowledge that it is preparing and presenting its important financial information in a manner that is both legal and ethical.

Sometimes it may not be clear whether an actual course of action is illegal or simply wrong. Put another way, an activity (including an accounting activity) may be legal, but still be the wrong thing to do. It is important that foodservice operators be able to make this ethical distinction.

Ethics refers to the choices of conduct made by an individual in relationships with others. Most individuals would agree that "ethical" behavior refers to behavior that is considered "right" or the "right thing to do."

Key Term

Ethics: A system or code of moral rules, principles, or values.

Consistently choosing ethical behavior is important to a foodservice operator's long-term career achievements. One reason: Sometimes an operator will not know what the law requires in a given situation. When managerial activities are examined, employers, in many cases, will simply consider whether an operator's actions were intentionally ethical or unethical.

How individual foodservice operators determine what constitutes ethical behavior can be influenced by their cultural or environmental background, religious views, professional training, and personal moral code. Also, the definition of ethical behavior may vary based upon an individual's own perception of what is ethical.

While it may sometimes be difficult to determine precisely what constitutes ethical behavior, the five guidelines in Figure 1.4 can prove useful when an operator evaluates ethical implications of a specific accounting-related decision or course of action.

Bookkeeping and accounting are the two processes used to record and analyze a foodservice operation's financial performance. Regardless of how foodservice operators accumulate, sort, and retain their financial data, outstanding operators use managerial accounting information as they compile and analyze that data.

In the next chapter, we will closely examine the standardized accounting procedures used to create the important summary financial documents used by foodservice operators to evaluate alternatives to help them better understand and manage their businesses.

---

1) **Is it legal?**

Any course of action that violates written law or company policies and procedures is wrong.

2) **Does it hurt anyone?**

Are benefits accruing to the operator that rightfully belong to the owner of the business? Discounts, rebates, and free products are the property of the business, not the manager.

3) **Am I being honest?**

Is the activity one that can comfortably reflect well on an operator's integrity as a professional, or will the activity actually diminish their reputation?

4) **Would I care if it happened to me?**

If an operator were the owner of the business, would he or she be in favor of their own managers behaving in the manner under consideration? If the operator owned multiple units, would it be good for the business if all unit managers followed the same courses of action?

5) **Would I publicize my action?**

Perhaps the most important guideline is that of considering whether the action to be taken could be publicized. A quick way to review the ethical merit of a situation is to consider whom an operator would tell about it. If the operator is comfortable telling their boss about the considered course of action, it is likely ethical.

If the operator would prefer that their actions go undetected, they are probably on shaky ethical ground. If an operator would not want their action to be read aloud in a court of law (even if the action is legal), then it likely should not be done.

Figure 1.4 Ethical Guidelines

What Would You Do? 1.2

Demario Davis was about to achieve his dream. Having finished his hospitality degree, Demario was ready to seek the financial backing he needed to open what he believed would be his first, but only the first, successful custom bake shop.

Demario was an experienced baker. With his educational background and solid industry experience behind him, Demario knew he had what it took to succeed.

To begin his project, Demario scheduled an appointment with the Small Business Administration (SBA) representative in his local area. Demario was meeting with the SBA to seek a federally backed loan that he needs to start his business. The SBA had asked him to bring his financial projections for the bake shop.

Assume you were the SBA representative meeting with Demario. How important would it be to you that Demario is knowledgeable and used GAAP to prepare the financial documents needed to apply for his loan? What would be your likely response if Demario did not understand and use such accounting principles?

Key Terms

Accounting
Accountant
Guest check average
Bookkeeping
Financial accounting
Auditing
Auditor
Tax accounting
Cost accounting
Managerial accounting
Uniform system of
    accounts

Uniform System of
    Accounts for
    Restaurants (USAR)
Generally Accepted
    Accounting
    Principles (GAAP)
Assets
Liability
Current fair value
Historical cost
Fiscal year
Calendar year

Accounting period
Expense
Revenue
Accrual accounting
    system
Accounts receivable (AR)
Accounts payable (AP)
Point-of-sale system
    (POS)
Transaction
Ethics

Operator's 10-Point Tactics for Success Checklist

Evaluate your need for, and the current status of, each of the following operational tactics. For those tactics you think are important, but not yet in place, develop an action plan for its implementation including who will be responsible for the tactic's completion and the target date by which it should be completed.

Tactic	Don't Agree (Not Done)	Agree (Done)	Agree (Not Done)	If Not Done Who Is Responsible?	Target Completion Date
1) Operator recognizes that the purpose of professional accounting is the proper reporting of a business's financial results.	——	——	——		
2) Operator understands the key differences between accounting and management.	——	——	——		
3) Operator recognizes the main purpose of each accounting specialization.	——	——	——		
4) Operator knows the purpose of a uniform system of accounts when reporting the financial results of a business.	——	——	——		
5) Operator understands the advantages of utilizing the Uniform System of Accounts for Restaurants (USAR) when reporting financial results in a foodservice operation.	——	——	——		
6) Operator recognizes the need to use GAAP in the preparation of a business's financial records.	——	——	——		
7) Operator understands the important role of bookkeeping in the financial management of a foodservice operation.	——	——	——		

(Continued)

Tactic	Don't Agree (Not Done)	Agree (Done)	Agree (Not Done)	If Not Done	
				Who Is Responsible?	Target Completion Date
8) Operator understands the important role of summary accounting in the financial management of a foodservice operation.	—	—	—		
9) Operator understands the important role of analysis in the financial management of a foodservice operation.	—	—	—		
10) Operator understands the importance of ethical behavior when recording and summarizing a foodservice operation's financial performance.	—	—	—		

2

The Mechanics of Accounting

What You Will Learn

1) The Basic Accounting Equation
2) How to Record Changes to the Basic Accounting Equation
3) The Importance of Financial Data Safety and Security

Operator's Brief

In this chapter, you will be introduced to the basic accounting equation that describes what a foodservice operation owns compared to what it owes. That **basic accounting equation** is:

$$Assets = Liabilities + Owners' Equity$$

A business's assets consist of many things including cash balances in banks, equipment and furnishings used in the business, buildings, and land. Liabilities include the debts a business owes to its vendors, suppliers, and others. Owners' equity is the amount of money business owners have invested in it, plus any profits made by the business, and minus any losses incurred by the business.

As bookkeepers and accountants record the financial transactions of a business, they must do so in a way that always keeps the basic accounting equation in balance. In most cases, foodservice operators use double-entry accounting to ensure the recording of each financial transaction is accurate.

The use of "T" accounts makes it easy to record changes in assets, liabilities, and owners' equity accounts. You will learn why and how this is done as you

Key Term

Basic accounting equation: A formula that shows that a company's total assets are equal to the sum of its liabilities and its shareholders' equity.

(Continued)

examine the concept of debiting and crediting "T" accounts to keep them in balance.

The accounting cycle is the entire process of recording all bookkeeping entries that affect the basic accounting equation and then preparing financial statements and summaries based on these entries. When performed properly, foodservice operators can analyze their current performance and make improvements to future performance based on their observations, experience, and insight.

In the concluding section of this chapter, you will be introduced to and review the importance of keeping a business's financial data safe and secure, a topic of increasing importance to all foodservice operators.

CHAPTER OUTLINE

Recording Business Transactions
The Basic Accounting Equation
Accounts Used in the Basic Accounting Equation
    Asset Accounts
    Liability Accounts
    Owners' Equity Accounts
    Revenue and Expense Accounts
Recording Changes to the Basic Accounting Equation
    Double-Entry Accounting
    The Journal and General Ledger
    Credits and Debits
Security of Financial Data

Recording Business Transactions

Large numbers of business transactions occur in foodservice operations. There is an exchange of money as guests use credit cards, electronic wallets, or cash to pay for the menu items they have purchased. Each guest making a purchase generates a financial transaction that must be properly recorded. Foodservice operators also make business transactions as they purchase food and beverage products, employee labor, and other resources. Bookkeepers record these transactions, and each transaction must be carefully analyzed to determine its effect on the business. After this analysis, the transaction must be properly recorded in the summary accounting records of the foodservice operation.

Business transactions typically impact the cash reserves held by a business. Guest purchases typically result in cash reserve increases, and business purchases yield decreases in cash reserves. However, some events occur that change the value of a business that do not involve transactions.

For example, as a piece of kitchen equipment wears out, periodic financial adjustments to its value called **depreciation** are made. Likewise, accounts receivable (AR) that cannot be collected require accounting adjustments for bad (uncollectable) debt expense. In all cases, the job of the bookkeeper or accountant is to record transactions and other events that result in changes to the "value" of the business.

Key Term

Depreciation: The allocation of the cost of equipment and other depreciable assets based on the projected length of their useful life.

Find Out More

Bookkeeping and accounting are similar activities, but they are not the same. Bookkeeping addresses recording and categorizing a business's daily income and expenses. It also includes conducting regular bank reconciliations and generating monthly financial statements.

In contrast, accounting is more forward-looking. For example, accountants analyze the costs of operating a business and use that information to make financial decisions. An important task of managerial accountants is to review and analyze the financial statements to ensure proper bookkeeping has been used. Managerial accountants also assess a business's financial health and make financial forecasts.

To learn more about the differences and similarities between bookkeeping and accounting tasks, enter "differences between bookkeeping and accounting" in your favorite search engine and review the results.

The Basic Accounting Equation

Everything "owned" by a foodservice operation are called its assets (see Chapter 1), and these commonly include cash, accounts receivable, equipment, land, buildings, food and beverage inventories, and various investments. Some assets are provided to the operation by its owners, while other assets might be obtained by borrowing money from a bank or other lenders.

There are, then, two groups who have claims to a foodservice operation's assets: owners and lenders. Broadly speaking, all claims to assets are called **equities,** and an operation's assets will always equal its equities.

Key Term

Equities: The claims against a foodservice operation's assets by those who provided the assets.

Equities are commonly divided into two groups:

1) Owners' Equity

Ownership claims to assets are called **owners' equity** if a foodservice operation is unincorporated. If the operation is a corporation, the term **stockholders' equity** is used.

2) Liabilities

The asset claims of outside parties such as financial institutions and suppliers are referred to as liabilities (see Chapter 1). The sum of owners' equities (internal equity) and liabilities (external parties' equity) will always equal a business's assets.

Key Term

Owners' equity: The assets of an operation minus its liabilities. Also, the net financial interest of an operation's owner(s).

Key Term

Stockholders' equity: A claim to assets of a corporately owned foodservice operation by the corporation's owners (stockholders).

A business's assets, liabilities, and owners' equity are all recorded in a basic accounting equation:

Assets = Liabilities − Owners' Equity

If the debts (liabilities) of a foodservice operation are subtracted from its assets, the result equals owners' equity. Liabilities represent the first claim to assets because, in an ongoing business, these must be paid regardless of their impact on owners' equity. When they are subtracted from assets, owners' equity (the residual claims to assets) remains. The basic accounting equation can then be rearranged to show this:

Assets − Liabilities = Owners' Equity

A basic financial statement (a point in time summary called the Balance Sheet) can be created to reflect its own basic accounting equation. When this is done, assets must equal (balance against) the liabilities and owners' equity.

An ability to read and understand balance sheets is an essential skill for foodservice owners and operators, and this important topic will be detailed in Chapter 4 of this book.

What Would You Do? 2.1

"I don't get it," said Vibiana, "are you telling me that we sold over seven thousand dollars' worth of gift cards for the holidays, but we didn't make any sales?"

Vibiana was the assistant manager at the Logjam Restaurant, a casual operation featuring grilled chicken items and a variety of comfort food–style homemade sides.

Vibiana was talking to Zendaya, the restaurant's manager, about a special marketing promotion they had run over the holidays. The restaurant had advertised $100 gift cards on sale for only $75. Vibiana had overseen marketing the promotion, and she had achieved great success: more than 100 gift cards had been sold in only two weeks.

"Well, actually they really weren't so much like a sale," replied Zendaya, "they were really more like a debt."

"So, are you saying we sold more than $7,000 worth of gift cards, and now we are in debt?" said Vibiana. "That doesn't make any sense!"

Assume you were Zendaya. How would you explain to Vibiana that a gift card sale generates an addition to their operation's liabilities, and the value of the unused cards is a liability on the balance sheet that is equal to the cash the operation received for the cards' sale?

Accounts Used in the Basic Accounting Equation

There is a very specific and unchanging relationship between a business's assets, liabilities, and owners' equity. Foodservice accountants can modify the equation stating this relationship by utilizing a variety of accounts. Note: An **account** is a device accountants use to record increases or decreases in the assets, liabilities, or owners' equity portions of a business.

Key Term

Account: An accounting device that shows increases and decreases in a single asset, liability, or owners' equity item.

It is important to recognize that the relationship between assets, liabilities, and owners' equity is expressed in a mathematical equation because it is so precise and unchanging. By using basic algebra, however, some useful and equivalent variations of the basic accounting equation can be created. These are presented in Figure 2.1.

One easy way to remember the rules that govern formula substitutions related to the basic accounting equation is to assign simple number values to each of the formula's three parts.

Assets = Liabilities + Owners' Equity

Assets – Liabilities = Owners' Equity

Assets – Owners' Equity = Liabilities

Figure 2.1 Variations of the Basic Accounting Equation

For example, if the number 10 is assigned to assets, the number 6 to liabilities, and the number 4 to owners' equity, the basic accounting equation becomes:

10 (Assets) = 6 (Liabilities) + 4 (Owners' Equity)

Or

10 = 6 + 4

Then, the variation of

Assets − Liabilities = Owners' Equity

becomes

10 − 6 = 4

The variation of

Assets − Owners' Equity = Liabilities

becomes

10 − 4 = 6

The important thing for operators to remember is that the basic accounting equation must always remain in balance. This will occur if each of a foodservice operation's individual financial transactions is entered correctly and into the appropriate account.

Asset Accounts

When foodservice operators record the value of their assets, they are recording information about several different asset types. Therefore, foodservice accountants create multiple accounts when they list an operation's assets. For most operators, among the most important of these asset accounts are:

Cash on hand This account includes house accounts such as petty cash and money in cash register banks.

Cash on deposit This account is the operation's bank account balance. If more than one bank account is used, for example, if one bank account is used for cash deposits and another for payroll disbursements, then separate cash on deposit accounts should be maintained for each of these assets.

Accounts receivable This account is used to record amounts due from guests that have not yet been received by the foodservice operation. This can occur, for example, when a foodservice operation hosts a large party. At the conclusion of the party, the restaurant sends an invoice to the guest who booked the party. Accounts receivable are also used to record regular purchases made by guests who have been extended credit by an operation.

Allowance for doubtful accounts This account provides a reserve for possible losses if/when some of an operation's accounts receivable are considered uncollectable.

Inventories A separate account should be maintained for each type of inventory. Food inventory consists of the cost of food on hand in food storage areas, pantries, kitchens, refrigerators, and freezers. Beverage inventory consists of all stock in bars and in the beverage storeroom. Separate inventory accounts should also be maintained for merchandise for sale (such as logoed hats, mugs, and T-shirts) and for cleaning, office, paper, and other supplies.

Prepaid expenses A separate asset account should be established for each prepaid expense item such as rent, licenses, and unexpired insurance.

Land This account is used to record purchases of land used in the business.

Building This account is used to record the purchase of buildings used in the business.

Equipment This account is used to record the purchase of equipment.

Furniture This account is used to record the purchase of interior and exterior furnishings.

China, glassware, silver, and linen This account is used to record the purchase of china, glassware, silver, and linen. In some cases, minor purchases of these types are expensed when they are purchased.

Accumulated depreciation This account is used for recording depreciation over the useful life of an asset such as a piece of kitchen equipment.

Liability Accounts

As with asset accounts, foodservice accountants create multiple individual accounts when they list an operation's liabilities. For most operators, among the most important of these liability accounts are:

Accounts payable—trade This account is used to record the amounts due to vendors and suppliers of goods and services in the foodservice operation's ordinary course of business. These are short-term (not long-term) debts.

Accounts payable—others This account is used to record extraordinarily large open accounts that might result from a major equipment purchase.

Taxes payable This group of accounts (one for each type of tax) is established to record taxes due to government authorities. Examples include federal, state, and city withholding taxes payables, FICA (Social Security) payable, sales taxes payable, and federal and state income taxes payable.

Deposits on banquets This account is used for recording deposits made by guests for future banquets/parties.

Gift certificates outstanding In this account, gift card sales are initially recorded as a liability because they indicate assets to be utilized in the future. They are recorded as sales after the card holders use the fund amount indicated on the card.

Accrued expenses This group of accounts is maintained for recording the amounts that are payable for expenses incurred at or near the end of an accounting period including accrued payroll, utilities, interest, and rent.

Dividends payable This account is for recording dividends payable based on formal declaration of dividend action by corporation's board of directors. This account is not applicable for those businesses that are unincorporated.

Long-term debt This group of accounts is used to record debt that is not due for 12 months from the balance sheet date. Examples include mortgage payable, notes payable, and bonds payable. A separate account is established for each long-term debt.

Owners' Equity Accounts

The precise accounting methods used to record owners' equity accounts varies somewhat based on whether a business is a sole proprietorship, a partnership, or a corporation. In general, however, owners' equity accounts include two major subcategories called permanent accounts and temporary accounts.

Permanent owners' equity accounts include items such as stock (or owners' investment) and **retained earnings**, the accumulated amounts of profits (or losses) over the life of the business that have not been distributed as **dividends**.

The specific types of permanent equity accounts maintained by a foodservice operation are dependent upon its type of ownership. The most important of these include:

✓ **For sole owner (proprietorships) or multiple owners (partnerships)**

(*Name of Proprietor or Partnership*) Capital This account shows the owner's net worth in a foodservice operation. The initial investment less withdrawals and operating losses plus operating profits results in this account's balance. If the business is organized as a partnership, a separate account is maintained for each partner.

Key Term

Permanent owners' equity account: An owners' equity account in which the balance at the end of an accounting period becomes the beginning balance for the next accounting period. Sometimes referred to as "real" accounts.

Key Term

Retained earnings: The cumulative net (retained) earnings or profits of a company after accounting for any dividends that have been paid out. Retained earnings decrease when a company loses money or pays out dividends, and they increase when new profits are generated.

Key Term

Dividends: A reward paid to the shareholders of a company for their investment in the company's equity. Dividends usually originate from a company's net profits.

✓ **For corporations**

Capital stock This group of accounts is used for recording each type of stock issued.

Paid-in capital in excess of par This account is used for recording proceeds from the sale of capital stock in excess of its **par value**. A separate account should be established for each type of stock issued.

Retained earnings This account records the amount of profits (earnings) retained (not returned to owners) in the foodservice operation.

Temporary owners' equity accounts include revenue and expense accounts. These temporary accounts can increase owners' equity (revenue accounts) or decrease owners' equity (expense accounts). They are used to show changes in the owners' equity account during a single accounting period.

At the end of an accounting period, temporary owners' equity accounts (primarily revenue and expense accounts) are **closed out,** and their balances are reduced to 0.

When closing temporary owners' equity accounts, a foodservice operation's current accounting period's net profit or loss is used to update the balance of the permanent owners' equity account (retained earnings).

Revenue and Expense Accounts

All revenue and expense accounts are actually temporary owners' equity accounts because they are closed out at the end of each accounting period to an operation's permanent owners' equity account(s).

Key Term

Par value (stock): The value of a single common share of stock as set by a corporation's charter. It is not typically related to the actual value of the shares. In fact, it is most often lower.

Key Term

Temporary owners' equity account: This includes income statement accounts (revenues, expenses, gains, and losses), the owner's drawing account (used to record the amounts withdrawn from a sole proprietorship by its owner), and the income summary accounts. These are temporary owner's equity accounts because, at the end of the year, the balances in these accounts are transferred to the owner's permanent capital account(s).

Key Term

Closed out (account): The accounting steps used to transfer amounts from temporary accounts to permanent accounts.

Accurate accounting records for revenue and expense accounts are extremely important for collecting and analyzing financial information used to make management decisions. The Uniform System of Accounts for Restaurants (USAR) (see Chapter 1) lists hundreds of possible accounts that could be utilized for recording

Revenue Accounts

Food sales	Other sales	Dividend income
Beverage sales	Interest income	

Expense Accounts

Cost of food sales	Uniforms	Cleaning supplies
Cost of beverage sales	Laundry	Packaging costs
Payroll	Linen rental	Guest supplies
Employee benefits	China and glassware	Menus
Payroll taxes	Silverware	Provision for doubtful accounts
Contract cleaning	Electricity	Repair expense
Flowers and decorations	Fuel	Insurance
Auto expense	Water	Professional fees
Licensing and permits	Waste removal	Delivery fees
Professional entertainers	Office supplies	Franchise fees
Print advertising	Postage	Interest expense
Online advertising	Telephone costs	Depreciation
Outdoor signs	Data processing costs	Amortization
Music service fees	Kitchen utensils	Income taxes

Figure 2.2 Common Foodservice Revenue and Expense Accounts

revenue and expenses. Most revenue and expense accounts are generally self-explanatory, and some of the most important of these accounts are shown in Figure 2.2.

The actual expense accounts utilized in accounting for a foodservice operation depends upon the specific needs of that operation. For example, a foodservice operation doing business in the mountains of Colorado will certainly require an expense account titled "Snow removal cost." A similar restaurant operating in Miami, FL, however, would not have the need for this expense account. Foodservice operators should review the choice of expense accounts listed in the USAR and select those most appropriate for their own operations.

The impact of entries made in permanent and temporary owners' equity accounts are shown in the following modification of the accounting equation:

Assets = Liabilities

+ Permanent owners' equity (Stocks + Retained earnings)

+ Temporary owners' equity (Revenue – Expenses)

In Chapter 1, it was stated that the use of generally accepted accounting principles (GAAP) requires the production of three major summary financial statements (income statement, balance sheet, and statement of cash flows). Two of these financial statements, the **balance sheet** and the **income statement**, are developed directly from the accounting equation.

The balance sheet (see Chapter 4) is an accounting summary that presents the financial condition (financial health) of a business. It does so by reporting the value of a company's total assets, liabilities, and owners' equity on a specified date.

The income statement (see Chapter 3) precisely reports, for a specific time period, a business's revenues from all its revenue-producing sources, the expenses required to generate those revenues, and the business's resulting profits or losses, which are often referred to as the operation's **net income**.

Recording Changes to the Basic Accounting Equation

Every time a business makes a financial transaction, it affects the basic accounting equation. Using the appropriate methods, foodservice accountants accurately and skillfully report all changes to the basic accounting equation while ensuring that the equation always remains in balance.

Consider Figure 2.3, a graphical representation of the basic accounting equation.

Conceptually, additions to or subtractions from one of the sides of the scale *must* be counterbalanced with an equal addition to or subtraction from the other side of the scale if the equation is to stay in balance. As a result, the scale will also remain in its required "equal" position if two entries are made (e.g., adding an equal amount or subtracting an equal amount from both sides).

Key Term

Balance sheet: A report that documents the assets, liabilities, and net worth (owners' equity) of a foodservice business at a single point in time. Also commonly called the Statement of Financial Position.

Key Term

Income statement: Formally known as "The Statement of Income and Expense," a report summarizing a foodservice operation's profitability including details regarding revenue, expenses, and profit (or loss) incurred during a specific accounting period. Also commonly called the Profit and Loss (P&L) statement.

Key Term

Net income: A number calculated as revenue minus operating expenses including depreciation, interest, and taxes, among others. It is useful to assess how much revenue exceeds the expenses of operating a business in a defined time period.

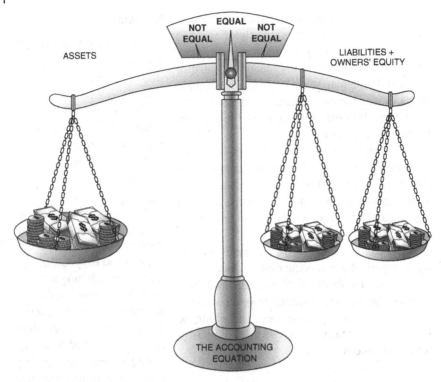

Figure 2.3 Graphical Representation of the Basic Accounting Equation

It is also possible to make changes (additions or subtractions) to only one side of the scale. For example, an equal dollar value added to and then subtracted from the asset total would not cause the overall equation to be out of balance. The accounting method used to help ensure that the left and right sides of the basic accounting equation always remain equal is known as **double-entry accounting**.

Double-Entry Accounting

It is important to keep accurate financial records, and this requires businesses in the United States and many other parts of the world to use the double-entry accounting system. As noted above,

Key Term

Double-entry accounting: A bookkeeping and accounting system that requires the recording of each business transaction in at least two accounts. While its use is not mandatory for private businesses, all public companies (those that issue stock) are required to use double-entry accounting to meet GAAP requirements. Also commonly called double-entry bookkeeping.

double-entry accounting requires that the person recording a financial transaction make at least two separate accounting entries (additions or subtractions) each time a financial transaction modifies the basic accounting equation.

The double-entry accounting system originated in Medieval Europe. According to most historians, the system was first devised and used extensively by Venice (Italy) merchants in the mid-1400s. A double-entry system helps to catch recording errors and to accurately track the various streams of money in and out of businesses. This is done not only for managers of the business but because others who need to know about a business's finances require information in a format they understand.

Today, more than 600 years after it originated, the agreed language of accounting is still double-entry accounting. For many foodservice operators, learning this language (just like learning to speak and understand any foreign language) can be confusing. It is not the intent of this book to make readers completely "fluent" in the language of double-entry accounting (doing so would likely require several college-level accounting courses). Instead, this book is intended to give foodservice operators a basic understanding of how the language of accounting is structured and how it is designed to operate.

The Journal and General Ledger

To better understand the complete accounting process, let's look at a foodservice operation. The operation's owner purchased required raw menu ingredients; hired staff; secured land, building, and equipment; and created and produced the menu items purchased by guests. As the operator did so, each individual financial transaction has a direct effect on the basic accounting equation and must be accurately recorded.

The double-entry accounting system requires each individual transaction to be recorded twice in the business's unique **journal** of financial transactions.

The journal is the written record of a specific operation's financial transactions. It would be possible to maintain two separate journals as a way of double-checking the accuracy of the business's financial records. However, maintaining only one journal and utilizing two entries per transaction for cross-checking purposes is an easier, more accurate, and convenient method than maintaining two separate records.

> **Key Term**
>
> **Journal:** A book for original entry of financial information into the accounting system.

A **journal entry** is made to a specific asset, liability, or owners' equity account when changes to the basic accounting equation are to be recorded. "Journalizing" is the accounting term that describes the procedure of recording transactions in the journal, and that is where transactions are analyzed, classified, and recorded.

The process of analyzing addressed above determines what type of transaction is to be recorded and the amounts that are involved. Classifying involves determining which accounts are affected, and recording is simply writing, processing, or in some other way entering information into the journal. The most simple and common journal is the **general journal**. All adjustments and closing entries for a business are recorded in its general journal.

The **general ledger**, also known as the book of *second* entry, is used to track assets, liabilities, owners' equity, revenues, and expenses. It is a book or file used to identify the amounts recorded in all relevant business accounts.

While foodservice operators may not be familiar with all of the details and nuances of proper accounting techniques, Figure 2.4 summarizes some of the important concepts they must recognize *prior to* learning about the ways managerial accountants actually record entries that will affect the basic accounting equation.

Key Term

Journal entry: The first step in the accounting cycle, and it is a record of the financial transaction in the accounting books of a business. A properly documented journal entry consists of the correct date, amounts entered, description of the transaction, and a unique reference number.

Key Term

General journal: Also referred to simply as the "journal," this accounting document is involved in the first phase of accounting because all transactions are initially recorded in it, originally and in chronological order. General journals are also often referred to as "the journal," an "individual journal," or "the book of original entry."

Key Term

General ledger: The main or primary accounting record of a business.

Credits and Debits

As stated earlier in this chapter, foodservice operators should use a double-entry accounting system to record each of their business financial transactions twice. Doing so helps minimize the chance of making a recording error and helps ensure that the accounting equation always stays in balance. Also, the asset, liability, and owners' equity portions of their business's individual basic accounting equation are broken down into smaller units called accounts.

For example, the worth of a foodservice operation's assets could be subdivided into separate and individual accounts that place a specific value on its cash bank

Important concepts for foodservice operators to remember about maintaining a business's financial records are:

1) The basic accounting equation that summarizes a business's asset, liability, and owners' equity accounts must always stay in balance.
2) The basic accounting equation is affected every time a business makes a financial transaction.
3) Each financial transaction should be recorded twice in a double-entry accounting system to reduce the chance of a recording error.
4) The original records of a business's financial transactions are maintained in its journal, and each financial transaction recorded is called a journal entry.
5) The current balances of each of a business's individual asset, liability, and owners' equity accounts are totaled and maintained in its general ledger.

Figure 2.4 Foundational Accounting Concepts

balance, food inventory, building, and land. Because of their shape, accountants often call these individual accounts **"T" accounts.**

Figure 2.5 shows an example of a "T" account. As can be seen in Figure 2.5, a "T" account consists of three main parts.

Part 1: The top of the "T" is used for identifying the name of the account. For example, a business would likely have a "T" account for "Cash" to identify that portion of its assets. If appropriate, it would also likely have a "T" account for "Loan" to identify, for example, that portion of its liabilities. A "T" account is created whenever an accountant wishes to add additional details to a portion of the accounting equation.

Part 2: The left side of the "T" account is called the **debit** side. Each journal entry made on the *left* side of the "T" account is always called a debit entry.

Key Term

"T" account: An informal term for a set of financial records that use double-entry bookkeeping. It is called a "T" account because the bookkeeping entries are laid out in a way that resembles a "T" shape. The title (name) of the account appears just above the "T."

Key Term

Debit (entry): Any entry made on the left side of a "T" account.

Name of "T" Account *(Part 1)*	
Left (Debit)	**Right (Credit)**
(Part 2)	*(Part 3)*

Figure 2.5 "T" Account

Part 3: The right side of the "T" account is called the **credit** side. Each journal entry made on the *right* side of the "T" account is always called a credit entry.

Note that the "T" account in Figure 2.5 looks very similar to the scale presented in Figure 2.3. In fact, the way an accountant uses "T" accounts to make a journal entry can be conceptualized just like the scale illustrating the accounting equation in Figure 2.3. The reasons are shown in the journal entry principles presented in Figure 2.6.

Foodservice accountants create individual "T" accounts to provide greater detail in financial reporting than could be achieved without them. For example, assume that the Regal Restaurant had $50,000 in cash in its bank account. Also assume that it owned a $5,000 convection oven located in its kitchen. Both items are clearly business assets of different types, and it is important that each type is accounted for separately.

Similarly, assume that the Regal Restaurant owed $8,000 per month on its 20-year mortgage (a liability). This is a different type of liability than the employment taxes owed by the restaurant (another liability), and for clarity it makes sense that they should be recorded in a separate "T" account.

Within each of the three major components of the basic accounting equation, accountants create individual "T" accounts to clarify the financial standing of a business. Some of the most commonly used individual accounts are shown in Figure 2.7.

Accumulated depreciation is a record of the accumulation of all depreciation expense charges that occur over the life of the asset. It is listed as a **contra asset** from the term "contra" meaning to deduct and represents deductions to the value of a fixed asset.

Key Term

Credit (entry): Any entry made on the right side of a "T" account.

Key Term

Accumulated depreciation: The sum of all recorded depreciation on an asset to a specific date.

Key Term

Contra asset (account): An account type used to decrease the balance in an asset account. The account is not classified as an asset since it does not represent a long-term value. It is not classified as a liability since it does not constitute a future obligation.

---

1) To make a complete journal entry, at least two different accounts must be used to record the event when using double-entry accounting.
2) Each journal entry must consist of at least one debit entry and one credit entry.
3) The total of all debit entries in a transaction must always equal the total of all credit entries.
4) When the above principles are followed, the accounting equation will always be in balance.
5) If the equation is not in balance, an error has been made in recording one or more journal entries, and it must be corrected.

Figure 2.6 Journal Entry Principles

Asset Accounts	Liability Accounts	Owners' Equity Accounts
CURRENT ASSETS	**CURRENT LIABILITIES**	**PERMANENT ACCOUNTS**
Cash	Accounts payable	Stock (or owners' investment)
Accounts receivable	Taxes due and payable	
Inventories	Notes payable	Retained earnings
FIXED ASSETS	**LONG-TERM DEBTS PAYABLE**	**TEMPORARY ACCOUNTS**
Furniture, Fixtures, and Equipment	Long-term loans	Revenue accounts
Buildings		Expense accounts
Land		
Accumulated depreciation (a contra asset account)*		

*As shown in Figure 2.7, accumulated depreciation is listed as a contra asset account (see discussion below). Recall that depreciation is a method of allocating the cost of a fixed asset over the useful life of the asset. Once fully depreciated, the value of the asset at the end of its useful life is called its **salvage value.**

Figure 2.7   Common Foodservice Operation "T" Accounts

Key Term

Salvage value: The estimated resale value of an asset at the end of its useful life. Salvage value is subtracted from the cost of a fixed asset to determine the amount of the asset cost to be depreciated so salvage value is a component of the depreciation calculation.

While many foodservice operators use various account titles suggested in the USAR, each individual operation should determine the most appropriate accounts for their own use. It is easy to understand this because "T" accounts are used to record changes to the basic accounting equation (each "T" account used will be modified when increases or decreases are made to its balance).

For example, funds in bank accounts (an asset "T" account) can increase or decrease, and money owed for employment taxes (a liability "T" account) may go up or down. Likewise, profits maintained by an operation's owners (an owners' equity "T" account) may increase or decrease.

Sometimes those who are new to double-entry accounting can make mistakes because it is easy to forget that a "debit" is made on the left side of a "T" account and a "credit" is made on the right side. It is easy to remember the correct way to make the entries, however, if one remembers that the word "debit" has one less letter in it than does the word "credit;" and it is also true that the word "left" has one less letter than the word "right!"

Five letters: Debit = Left
Six letters: Credit = Right

"T" accounts have common characteristics. For example, an asset account such as "Cash" will typically have a positive balance because the account will reflect money on hand. As a result, a "T" account set-up to record and monitor the value of this money will generally have a debit balance. Additions to the current balance of an asset account such as this one are traditionally recorded on the left (debit) side of a "T" account.

Reductions in the value of an asset account are traditionally recorded on the right (credit) side of its "T" account. The difference between a "T" account's total debits and total credits is called the **account balance**.

Figure 2.8 shows the normal or expected balances of various types of "T" accounts.

Key Term

Account balance: The total amount of money available in a financial account after all the debits and credits have been calculated.

Recall that each journal entry made on a "T" account affects its account balance. Figure 2.9 summarizes the impact of each kind of entry on each of the three major components of the basic accounting equation.

To further illustrate how the use of debits, credits, and double-entry accounting affects individual "T" accounts and the basic accounting equation, consider the Stagecoach Restaurant. It has just begun its operation with a $1,000,000 check from its owner. The two accounting transactions required to initiate the operation's accounting system are as follows:

Transaction 1: A $1,000,000 debit to "Cash" (↑ asset)
Transaction 2: A $1,000,000 credit of the owners' investment (↑ owners' equity)

"T" Account Type	Normal Balance
Asset	Debit
Liability	Credit
Owners' Equity	
Permanent Accounts (Owners' Equity)	Credit
Temporary Accounts	
Revenue	Credit
Expense	Debit

Figure 2.8   Normal Balances of "T" Accounts

Accounting Equation Component	Journal Entry Made	Impact on Balance
ASSETS	Debit	Increases
	Credit	Decreases
LIABILITIES	Debit	Decreases
	Credit	Increases
OWNERS' EQUITY		
Permanent Accounts (Owners' Equity)	Debit	Decreases
	Credit	Increases
Temporary Accounts		
Revenue	Debit	Decreases
	Credit	Increases
Expense	Debit	Increases
	Credit	Decreases

Figure 2.9 Impact of Debit and Credit Entries on Basic Accounting Equation Components

The resulting "T" accounts are established as follows:

Cash	
$1,000,000	

Owners' Equity	
	$1,000,000

Upon completion of these two initial journal entries, the Stagecoach Restaurant's basic accounting equation would be in balance and read as:

Assets = Liabilities + Owners' Equity

$1,000,000 = 0 + $1,000,000

Now assume that the operation's owner established a T account titled "Uniforms," in addition to the "Cash" T account previously established. The operator then purchases, with cash, $1,000 worth of uniforms for the operation's future dining room staff. This action *decreases* the amount of money in the "Cash" asset account and *increases* the value of the "Uniforms" asset account.

The two accounting transactions to be made are as follows:

Transaction 1: A $1,000 credit to "Cash" (↓ asset)
Transaction 2: A $1,000 debit to "Uniforms" (↑ asset)

The resulting "T" account entries are as follows:

Cash	
	$1,000

Uniforms	
$1,000	

Now assume further that the Stagecoach owner purchases a vacant lot adjacent to the restaurant to expand its parking area. The lot is purchased for $50,000, and the owner secures a bank loan to finance the purchase. The accounting transactions needed to record this activity are as follows:

Transaction 1: A $50,000 debit to "Land" (↑ asset)
Transaction 2: A $50,000 credit to "Loans Payable" (↑ liabilities)

The resulting "T" account entries are as follows:

Land	
$50,000	

Loans Payable	
	$50,000

The Stagecoach's basic accounting equation is now revised and reads:

$1,000,000$ Cash $+ \$50,000$ Land $= \$50,000$ Loans Payable $+ \$1,000,000$ Owners' Equity

Assets = Liabilities + Owners' Equity

Or

$$\$1,050,000 = \$1,050,000$$

At the conclusion of the land purchase, both sides of the basic accounting equation are still equal (in balance). Note that the $1,000 debit to uniforms and the $1,000 credit to cash addressed previously are not included in the equation because they are both current assets and cancel each other out. A foodservice operation's revenue and expense accounts are among the most important and frequently used "T" accounts. As noted earlier, these two account types belong to the owners' equity portion of the accounting equation and are summarized and closed out at the end of each accounting period.

For some foodservice operators, the principles that accountants use to record financial transactions are easy to understand; other operators may have more difficulty fully mastering them. All managers, however, can learn the specific recording procedures mandated at their own properties. For managerial accountants, the truly important concepts to remember are those of careful and accurate **posting** of all financial transactions and the responsibility to ensure that all accounts and, therefore, the basic accounting equation remain in balance.

Key Term

Posting: Moving a transaction entry from a journal to a general ledger.

Technology at Work

The demands of operating a foodservice business are such that few managers perform the daily bookkeeping and summary accounting tasks required to produce useful financial summaries of their businesses.

In many cases foodservice operators employ a professional accountant whose responsibilities include ensuring proper posting of income and expenses and the creation of summary income statements, balance sheets, and statements of cash flows. Some foodservice operators, however, utilize accounting software developed specifically for the foodservice industry. Note: "QuickBooks" is an example of a popular software program.

The QuickBooks product line includes several different programs that can be adapted for use by foodservice operators. To learn more about this accounting software program, enter "QuickBooks for restaurants" in your favorite search engine and review the results.

Regardless of their personal level of involvement in the bookkeeping and accounting activities of their businesses, all foodservice operators must have faith in the information generated by accounting systems. As a result, ensuring the use of accurate bookkeeping and accounting processes is critical. While foodservice operators need not be **CPAs** to manage their businesses, it is their responsibility to ensure their operations' recordkeeping efforts meet industry standards.

Security of Financial Data

Regardless of the specific methods used by foodservice operators to record and analyze their accounting information, any serious examination of accounting data in today's business world must address the important issues of **data privacy** and **data security**.

Data privacy addresses the issues of authorized data collection, utilization, and sharing. Data security includes actions necessary to ensure data is not accessed by those who might steal or manipulate it.

Keeping accounting data secure is especially important considering the sensitive nature of the information foodservice operators routinely gather and maintain. Figure 2.10 lists some of the guest- and property-related information typically stored in a foodservice operation's accounting records that must be protected from unauthorized users.

Increasingly, sophisticated hackers attempt to gain access to foodservice operators' internal data for unscrupulous purposes. These hackers use one or more phishing, smishing, or vishing schemes to do so.

Phishing is a method of cyberattack that tricks victims into clicking on fraudulent links sent in e-mails.

Key Term

CPA: Short for Certified Public Accountant. A CPA is a professional designation given to qualified accountants who have passed a rigorous accounting exam (the Uniform CPA Exam).

Key Term

Data privacy: The aspect of data management that addresses who has ownership over data, who can determine its legitimate use, and the regulations related to utilizing this data.

Key Term

Data security: The steps taken and the safeguards implemented to prevent unauthorized access to an organization's data.

Key Term

Phishing: The fraudulent practice of sending e-mails supposedly from reputable companies to induce individuals to reveal personal information including passwords and credit card numbers.

A phishing link sent in an e-mail typically takes the victim to a seemingly legitimate form that requests them to provide usernames, passwords, account numbers, or other private information then sent directly to the cyberattackers.

For example, a foodservice operator may receive an e-mail stating that their operation's bank account has been locked to protect it against theft. The operator is

Guest-Related Data	Property-Related Data
Names	Revenue histories
Addresses	Sources of business records
Phone numbers	Names of loyalty club members
E-mail addresses	Banquet contract–related information
Payment card information	Payroll data
Purchase histories	Operating cost totals
Outstanding bills	Profit levels

Figure 2.10 Common Guest-Related and Property-Related Accounting Data Generated by Foodservice Operations

requested to click on a link to open and regain access to the account. The link leads to a fraudulent form requesting banking information including one's online banking username and password. Using this information, the cyberattackers can then log in to the operation's bank account and steal funds available.

Smishing is a type of cyberattack like phishing, but it comes in the form of a text message.

Like a phishing attack, a smishing text often contains a fraudulent link that takes its victims to a form used to steal their information. The link may also be used by the cyberattackers to download **malware** onto the victim's device.

Typically, the smishing text messages may appear to be urgent requests sent from a bank or vendor. The e-mail may claim there has been a large withdrawal from a foodservice operator's bank account, or that an unknown vendor is confirming a very large order to be delivered immediately. Foodservice operators can fall for this scam if they believe they must take quick action to solve an urgent problem.

Fraudulent calls or voicemails fall under the category of "**vishing**." When using vishing scams, cyberattackers call their potential victims, often using prerecorded robocalls, and pretend to be a legitimate company to solicit personal information from an intended victim.

Key Term

Smishing (attack): The fraudulent practice of sending text messages supposedly from reputable companies to induce individuals to reveal personal information such as passwords or credit card numbers.

Key Term

Malware: Software that is specifically designed to disrupt, damage, or gain unauthorized access to a computer system.

Key Term

Vishing: The fraudulent practice of making phone calls or leaving voice messages purporting to be from reputable companies to induce individuals to reveal personal information such as bank details and credit card numbers.

In a vishing call, the receiver may be asked to provide information such as their first and last name, address, driver's license number, Social Security number, or credit card information. Some attackers using the vishing scam may record the voice of the call's receiver and ask a question the receiver is likely to answer with a "Yes." The cyberattacker can then use this recording to impersonate the call's receiver to authorize charges or access financial accounts.

Regardless of the attack method used, operators must understand the significant problems and negative publicity that can arise when a business's data systems are hacked or compromised. Data breaches can happen within a foodservice operation itself, but they can also occur when an operation's third-party partners experience a data breach.

Foodservice operators supply food and beverages to customers, but they can also unknowingly supply credit card data to hackers. Given the volume of credit card transactions, foodservice operators must take cybersecurity seriously. A hacker only needs to gain access to a foodservice operator's POS system (see Chapter 1) and install malware to steal customer credit card details. This information can be used to sell stolen card data on the dark web or used as a springboard for identity theft scams.

A significant and highly publicized data breach can also cause guests to be wary of visiting a foodservice operation or even a chain of operations. Many customers stop doing business with operations that suffer a data breach. Most foodservice operations, and especially smaller ones, cannot afford to be lax on cybersecurity because each guest is important to their long-term financial viability.

As a result, all foodservice operators and their staff members must protect the data they create in a manner that is consistent with their operation's data privacy and security policies. They must also comply with emerging legislation that increasingly dictates how businesses must secure their customers' personal data.

Find Out More

Hackers continue to target the foodservice industry with sophisticated attacks on secured data. These attacks can affect even the best hospitality organizations. Here are just a few high-profile examples:

Who: Landry's Inc.   **When:** October 2019

What happened: Landry's, a Houston-based company with 600 restaurants, hotels, and casinos in the United States, warned customers about a data breach that could have compromised credit card information. The data breach likely affected payment cards used in the seven-month period between March and October 2019.

Who: Various Online Ordering Systems   **When:** January 2022

What happened: Three restaurant ordering platforms (MenuDrive, Harbortouch, and InTouchPOS) were the target of two Magecart skimming campaigns that resulted in the compromise of at least 311 restaurants. The attacks inserted malicious PHP code (PHP is an acronym for "PHP: Hypertext Preprocessor," a widely-used, open source scripting language) into the businesses' online check-out pages by using known security flaws in the services to scrape and transmit the customer data to a server under the attacker's control. The trio of breaches has led to the theft of more than 50,000 payment card records from these infected restaurants and data has been posted for sale on the dark web.*

Who: DoorDash   **When:** August 2022

What happened: DoorDash, one of the popular third-party delivery services reported its own data breach in August 2022 (its database was partially hacked via a phishing attack on one of its own a third-party vendors).**

There is little doubt that data safety and security will continue to be critical issues for foodservice operators. To review important current information related to the security of a foodservice operation's accounting data, enter "tips for ensuring restaurant data security" in your favorite search engine and review the results.

* *https://thehackernews.com/2022/07/magecart-hacks-online-food-ordering.html,* retrieved September 22, 2022.

** *https://www.cnet.com/tech/services-and-software/customer-information-stolen-in-doordash-data-breach/,* retrieved September 22, 2022.

Technology at Work

As increasing numbers of foodservice operators implement remote work, move important data to the cloud, and adopt new wireless ordering and pay-ment technology, threats to the safety and security of their accounting data also continue to increase.

Fortunately, there are companies that specialize in assisting foodservice operators in securing the important data generated by their businesses. To examine some cybersecurity firms available to help foodservice operators secure their most important financial information, enter "assistance in secur-ing restaurant operating data" in your favorite search engine and review the results.

This chapter has summarized basic accounting procedures foodservice operators utilize to accurately record their business transactions and make changes to the basic accounting equation applicable to their individual businesses. When done appropriately, summary financial reports can be produced to help operators better manage their businesses.

One of these important summary reports is the income statement (also known as a profit and loss statement, P&L, statement of operation, statement of financial result or income, or earnings statement). Regardless of its name, an ability to read and understand the information presented in this statement is essential for all foodservice operators. Learning how to read and interpret a foodservice operation's income statement is so important that it is the sole topic of the next chapter.

What Would You Do? 2.2

"Everybody is offering free Wi-Fi!" said Liam, "and our Internet service provider said our download speeds are so fast that having even a dozen customers on our network at the same time wouldn't cause us a problem. We just need to give them our system password."

Liam, the manager at the Irish Rose coffee shop, was talking to Roison Dubh, the shop's owner.

"I agree it would be good for customers, and it is a service we should offer," Roison replied, "but it's not as easy as just providing our guests with a password to our Wi-Fi server."

"Why not?" said Liam, "sometimes, when I am home, I log into our system on my laptop, and it never causes any problem," said Liam.

"That's just my point," replied Roisin, "if you know our system's name and password and can access our accounts from anywhere, so could a customer who knows the password to our Wi-Fi system. If a hacker has access to our network, they could access our router. And that means trouble."

"I don't know about all that computer stuff," said Liam, "all I know is free Wi-Fi will help us attract more customers."

Assume you are the owner of the Irish Rose coffee shop. Do you think offering free Wi-Fi access would be important to guests? How important is ensuring guests do not have access to the Wi-Fi network used by the business to maintain its sales data and customer payment-related information?

Key Terms

Basic accounting equation	Closed out (account)	Accumulated depreciation
Depreciation	Balance sheet	Contra asset (account)
Equities	Income statement	Account balance
Owners' equity	Net income	Posting
Stockholders' equity	Double-entry accounting	CPA
Account	Journal	Data privacy
Permanent owners' equity account	Journal entry	Data security
Retained earnings	General journal	Phishing
Dividends	General ledger	Smishing
Par value (stock)	"T" account	Vishing
Temporary owners' equity account	Debit (entry)	
	Credit (entry)	
	Salvage value	

Operator's 10-Point Tactics for Success Checklist

Evaluate your need for, and the current status of, each of the following operational tactics. For those tactics you think are important, but not yet in place, develop an action plan for its implementation including who will be responsible for the tactic's completion and the target date by which it should be completed.

				If Not Done	
Tactic	Don't Agree (Not Done)	Agree (Done)	Agree (Not Done)	Who Is Responsible?	Target Completion Date
1) Operator realizes the importance of accurately recording and reporting all the operation's business transactions.	——	——	——		
2) Operator understands the basic accounting equation: Assets = Liabilities + Owners' equity.	——	——	——		

(Continued)

Tactic	Don't Agree (Not Done)	Agree (Done)	Agree (Not Done)	If Not Done	
				Who Is Responsible?	Target Completion Date
3) Operator is familiar with the most common asset accounts utilized in the basic accounting equation.	——	——	——		
4) Operator is familiar with the most common liability accounts utilized in the basic accounting equation.	——	——	——		
5) Operator is familiar with the most common owners' equity accounts utilized in the basic accounting equation.	——	——	——		
6) Operator is familiar with the most common revenue and expense accounts utilized in the owners' equity portion of the basic accounting equation.	——	——	——		
7) Operator knows the advantages of using a double-entry accounting system.	——	——	——		
8) Operator understands the recordkeeping role of a business's journal.	——	——	——		
9) Operator understands the purpose of "T" accounts, debits, and credits as they relate to maintaining the accuracy of a business's general ledger.	——	——	——		
10) Operator appreciates the importance of maintaining the safety and security of all of their business's and customer's financial data.	——	——	——		

3

The Income Statement

What You Will Learn

1) The Importance of the Income Statement
2) How an Income Statement Is Formatted
3) How to Read and Analyze an Income Statement

Operator's Brief

In this chapter, you will learn why the income statement is so important to those who must read and analyze it. You will also learn why those who operate a business are able to read and correctly analyze and interpret it. Perhaps most importantly, you will learn how foodservice operators use their income statements to learn as much as possible about the current and future profitability of the business.

When you manage a foodservice operation, you will receive revenue (sales or the money taken in) and incur expenses (the cost of the items required to operate the business). The dollars that remain after all expenses have been paid represent your profit. All this information is summarized on an income statement and will be important to many including an operation's owners, investors, lenders, and creditors.

An income statement prepared according to the Uniform System of Accounts for Restaurants (USAR) lists an operation's sales (revenue) first. Next are listed the operation's total cost of sales (the cost of the ingredients used to prepare the menu items sold) and its total labor costs (including all management and staff costs, and employee benefit costs).

The total cost of sales and total labor costs are then subtracted from sales to yield the operation's prime costs. Other controllable expenses and

(Continued)

non-controllable expenses are then listed and, finally, the income statement shows the operation's income (profit or loss) before income taxes.

It will be very important for you to be able to read and analyze the major components of a USAR income statement so that you can better understand and manage your business in the future.

CHAPTER OUTLINE

The Importance of the Income Statement
    Users of the Income Statement
    Frequency of Income Statement Preparation
USAR Income Statement Format
    Sales (Revenue)
    Total Cost of Sales
    Total Labor
    Prime Costs
    Other Controllable Expenses
    Non-Controllable Expenses
    Profits (Income)
EBITDA
Utilizing the Income Statement
    Supporting Schedules
    Comparative Analysis

The Importance of the Income Statement

Because all businesses want to be profitable, foodservice owners and managers want their operations to make a profit. The tool operators use to document and report their profits is called the income statement (see Chapter 2). The income statement is formally known as "The Statement of Income and Expense." Some operators refer to the income statement as the **"profit and loss" (P&L) statement** and, while that name is very much in common use, in this book we will simply refer to the document by its shortened name: the income statement.

A foodservice operation's revenue minus its expenses equals its profits. The income statement details, for a specific time period, an operation's revenue from all revenue-producing sources, and it also reports the expenses required to generate those revenues and the resulting profits or losses.

Key Term

Profit and loss (P&L) statement: Formally known as "The Statement of Income and Expense," this is a report summarizing a foodservice operation's profitability including details about revenue, expenses, and profit (or loss) incurred during a specific accounting period. Also commonly referred to as the "income statement."

The following is the basic profit formula, and it is also the format followed when preparing an operation's income statement.

Revenue – Expense = Profit

Many operators use the following terms interchangeably: sales and revenue, expenses and costs, and profit and net income.

To illustrate the use of the formula, consider Joshua, the operator of Joshua's restaurant. Joshua's operation will generate revenue from a variety of sources. These revenue sources can include:

- ✓ In-house food sales
- ✓ In-house alcoholic beverage sales
- ✓ Gift card sales
- ✓ Carry-out sales
- ✓ On-site banquet (catering) sales
- ✓ Off-site banquet (catering) sales

In addition to generating revenue, Joshua generates expenses and incurs costs as he operates each of these major revenue centers. His goal should be to generate a profit in each of the revenue sources he manages. In fact, many foodservice operators consider each individual revenue-generating segment within their business to be a **profit center,** a term coined in 1945 by international management consultant Peter Drucker.[1]

The *Revenue – Expense = Profit* formula is also applicable to what is not typically considered a for-profit segment of the foodservice industry. For example, consider the operation managed by Hector Bentevina.

Hector is the manager of the employee cafeteria at Consolidated Industries, a large international corporation employing over 2,000 workers at its company headquarters. Hector provides employee meals at no cost to a large group of service workers and managerial staff working at the company.

In this situation, Hector's company clearly does not have making an immediate "profit" as its primary motive in operating the cafeteria. In fact, many **business dining** and other **non-commercial foodservice** settings provide food

Key Term

Profit center: A part of a business organization with assignable revenues and expenses. Also referred to as a "revenue center."

Key Term

Business dining: The non-commercial segment of the foodservice industry that serves the dining needs of large businesses and corporations.

Key Term

Non-commercial foodservice: A foodservice operation created primarily to support an organization's larger mission such as education, healthcare, or business/industry.

---

1 https://en.wikipedia.org/wiki/Profit_center#:~:text=Peter%20Drucker%20originally%20 coined%20the,cheque%20hasn't%20bounced%E2%80%9D, retrieved November 11, 2022.

as a service to employees, students, patients, military, or others as a no-cost or greatly reduced-price benefit. Therefore, in these situations, operators manage a **cost center** that generates expenses, but with little or no revenue paid by the meal consumers. Note: The managers of the cost center will have a contract that specifies, among other issues, payments to Hector along with the responsibilities he incurs for these payments.

Key Term

Cost center: A part of a business organization with assignable expenses that generates little or no revenue.

In the non-commercial segment of the foodservice industry, an operator's goal may be to maintain or not exceed a predetermined cost for operating the cost center as an employee (labor) benefit. Whether they are operating a profit center or a cost center, however, all foodservice operators must know as much as possible about how their revenues and expenses are generated to best maximize their income and control their costs. The income statement is the primary tool foodservice operators use to gather and assess this necessary revenue and expense information.

Find Out More

Non-commercial foodservice is a secondary support service in many educational institutions and other organizations. The non-commercial enterprise accounts for about 23% of food expenditures outside the home. Providing food and beverages is not the main goal of these operations. Rather a foodservice facility is offered as a secondary service to support an organization's main purpose.

For example, schools and nursing homes frequently offer non-commercial foodservices to, respectively, students and faculty and short- or long-term patients. These types of establishments are often called "institutional" foodservice facilities, and other non-commercial operations in the food industry include:

✓ Private clubs
✓ Hospitals
✓ Military
✓ Prisons
✓ Colleges/Universities
✓ Transport catering
✓ Industrial catering
✓ Institutional catering

In some cases, these operations generate revenues, but in other situations, they do not. Consider, for example, a company providing foodservices at a price lower than total costs: employees consider their meal savings as a "benefit," and the difference between actual and charged costs is considered a labor benefit by the company. As a result, foodservice operators working in the non-commercial segment of the industry must confront issues related to controlling their expenses just as much as do those operators who work in the commercial segment. In most cases, they will use an income statement as an important tool to do so.

To learn more about some of the other important issues non-commercial foodservice professionals must address, enter "challenges in operating a non-commercial foodservice" in your favorite search engine and review the results.

Users of the Income Statement

Those who operate a foodservice are not the only ones interested in the business's revenue, expenses, and profits. All **stakeholders** (see Chapter 2) who are affected by an operation's profitability will care greatly about the effective operation of a foodservice business. These stakeholders may include:

✓ Owners
✓ Investors
✓ Lenders
✓ Creditors
✓ Managers

Key Term

Stakeholder: A person or organization with a vested interest in a foodservice operation that can either affect or be affected by the operation's financial performance.

The owners and operators (and employees!) of a foodservice business all benefit when the business is successful. As a result, when an accurate income statement provides useful information, the business's owners, lenders, investors, and managers can all make better decisions about how best to operate it or even expand it.

Owners
The owners of a business typically have the greatest interest in its success because they often share in the profits generated by the business. In many foodservice operations, however, the owners do not actively participate in managing the business. Then, business owners cannot exercise close control because they are physically removed from the business. By evaluating income statement data, owners can better determine the effectiveness of manager(s) they have selected to operate their businesses, and the progress that has been made to achieve the owner's operating goals.

Investors

Investors supply funds to foodservice operators to earn money on their investments. These earnings generally include periodic cash payments from profits generated by the business, plus any property **appreciation** (increase in value) achieved during the period of the investment.

Commercial foodservice operations are inherently risky investments because they require specialized management expertise, can be subjected to economic upturns and downturns, and are often not easy to sell quickly. This may be especially true if the business is not operating profitably.

Before individuals, corporations, or other financial entities elect to invest in a business, they will want to know that their investment is a good one. Generally, the higher the **return on investment (ROI)** sought by an investor, the riskier is the investment. It is also true that different investors, with different investment goals, will evaluate and choose from a variety of investment options.

Key Term

Appreciation: An increase in the value of a business asset over time. Unlike depreciation, which lowers an asset's value over its useful life, appreciation is the rate at which an asset grows in value.

Key Term

Return on investment (ROI): A measure of the ability of an investment to generate income.

An investor's ROI is expressed as a percentage of money earned to money invested using the following formula:

$$\frac{\text{Money Earned on Funds Invested}}{\text{Funds Invested}} = \text{ROI}$$

For example, if an operator invested $1,000,000 in a foodservice business, and at the end of the year the business generated $200,000 in profits, the operator's ROI would be calculated as follows:

$$\frac{\$200,000}{\$1,000,000} = 0.20 \text{ or } 20\% \text{ ROI}$$

Stated another way, money earned on funds invested are equal to the amount of profits an investment has produced. Since the income statement lists a foodservice operation's profits (or losses!), it is a critical document in calculating an investor's ROI.

Not all foodservice operations produce profits immediately upon opening. When an investment produces losses, an investor's ROI will be negative. Few investors seek long-term negative returns on their investment, although investors recognize that the early years of an investment may yield negative returns. These investors are willing to accept negative returns in the short term in exchange for the belief they will achieve significant positive investment returns in the long term. A close examination of a foodservice operation's income statements allows investors to chart a new operation's progress toward profitability.

In most cases, those who invest in a foodservice operation require that the business regularly produce and supply accurate and timely income statements prepared according to GAAP (see Chapter 1). Typically, an income statement must be provided no less than 12 or 13 times (depending upon the length of a business's accounting period) per year with a fiscal year-ending income statement summarizing revenue, expense, and net income data for the business's full operating year.

Lenders

Many foodservice operations are financed with both debt (funds lent to a business) and equity (funds supplied by investors or owners). Those who lend money to a business generally require that the loan amount be repaid with applicable interest. Lenders to businesses in the foodservice industry include banks, insurance companies, pension funds, governmental agencies, and other similar capital sources. Lenders may or may not actually own a part of the business in which they put funds, but they do agree to provide funds for its construction and/or operation.

There are essentially two types of entities that lend money to foodservice operators. The first type are lenders who simply agree to fund or finance a business. Typically, this is done by granting the business a loan that must be repaid according to specific terms and conditions of the loan. Lenders to a business have first claim to the profits generated by the business. As a result, lenders will be repaid before a business is permitted to distribute profits to its investors, and the lender's ROI requirements are typically lower than those of investors.

Investors, however, do *not* know what their final ROI will be until the business in which they invest summarizes operational revenue and expenses in an income statement. In contrast, lenders typically establish the required ROI before they make a loan. For example, a bank may lend $1,000,000 to Joshua's Restaurant (see above). The loan's terms might state that the loan must be repaid along with 8% interest. Note: The interest rate represents the ROI sought by the bank's managers for their bank's money.

Additional loan terms likely include information about how much is to be repaid each month, when payments will be made, and any fees the lender charges the business for initiating and supplying the funds. Like investors, lenders seek greater returns (higher interest rates) when they perceive high risks that a business will not repay its loan.

Alternatively, when a lender believes the risk of default (non-payment) of a loan is low, interest rates charged for the use of the lender's funds will also be lower. To determine a business's ability to meet the repayment terms of loans offered to it, a lender will want to carefully review the income statements generated by the business.

Creditors

A **creditor** is a vendor, supplier, or service provider to whom a business owes money. For example, when the manager of Joshua's Restaurant must purchase food from a vendor, that vendor is likely to extend credit to the operation. The food

order is delivered, and an invoice detailing the purchases including their costs is also provided during the delivery.

The vendor's expectation is that the restaurant will pay the vendor for the delivered food according to the credit terms including the time agreed upon by the vendor and the operation. Until that payment is made, however, the creditor has, for all practical purposes, made a "loan" to the operation equal to the dollar value of the food or other products delivered.

Foodservice operators should understand that a vendor's business must be profitable, and vendors charge higher prices to those businesses they believe have a high risk of not paying according to the agreement. Conversely, customers with a lower risk of non-payment will often be charged lower prices, so vendors are very interested in the creditworthiness of their customers. Over the long run, a company's ability to pay its bills promptly is a result of its profitability. Profitability of a business is measured, in part, by the information contained in the income statement. Information from that document will be of interest to many vendors utilized by a business, and information from it may be required before the vendor grants credit to the business.

Managers

Many foodservice managers consider the income statement to be the best reflection of their managerial ability for several reasons. First, it details how profitable an operation has been within a designated time, and the best managers operate more profitable facilities than do poorer managers. Operational performance as measured by the income statement's results is often used to establish managers' pay raises, compute their bonuses, and performance is also important in determining promotional opportunities. Therefore, managers must be able to read and understand their business's income statements.

Owners, investors, lenders, vendors, and managers are not the only ones interested in the results of a business's income statement. Employees are also important stakeholders in a business. While non-management employees do not typically view the actual income statement of their employers, the information the income statement contains will have a large impact on the potential stability of their employment.

For example, if a business is profitable, employees' jobs are more secure than if the business is struggling. Similarly, employee wages are more likely to rise, business expansion may create additional job opportunities, and revised employee benefit programs will better satisfy employee and business needs.

Frequency of Income Statement Preparation

An income statement can be prepared for any accounting period determined by a foodservice operator (see Chapter 1). However, the accounting periods should make good sense for the business to which they are applied. For many businesses, accounting periods coincide with the calendar months of the year (calendar year accounting period) or any consecutive 12-month period (fiscal year accounting period). Many (but not all) businesses produce a calendar year income statement because it eases the operation's ability to file annual tax returns, which are often based upon the calendar year.

In addition to preparing their annual income statement, most foodservice operators also prepare shorter accounting period income statements. Sometimes, they address a four-month (quarterly) accounting period, but more often they are created monthly. In some cases, a foodservice operator may prefer to create income statements that are 28 days long because they want to create "equal" **28-day accounting periods**.

When preparing income statements based on a 28-day accounting period, each period is equal in length and has the same number of Mondays, Tuesdays, Wednesdays, and so forth. This helps the operator compare performance from one accounting period to the next without having to compensate for "extra days" in any one period.

Some foodservice operators also create weekly or daily (or even hourly!) "mini" income statements. These shortened documents frequently lack the specific and accurate revenue and (especially) the expense data needed to produce an income statement meeting standards including GAAP.

Figure 3.1 summarizes the most popular lengths of time used by foodservice operators and their accountants to create income statements.

Key Term

28-day accounting period:
An accounting period that is four weeks (28 days) in length instead of a calendar month that has between 28 and 31 days. There are 13 four-week periods instead of 12 monthly periods when using this system.

Accounting Period	Number of Days Included
Month	Varies
Quarter	Varies
Annual	365 days (except leap year)
28-day	28 days
Weekly	7 days
Daily	1 day

Figure 3.1 Common Income Statement Accounting Periods

USAR Income Statement Format

There are a variety of ways that foodservice operators *could* report their revenue, expense, and profits. Chapter 1 of this book introduced the USAR. The USAR has recommendations for how an income statement *should* be produced. The USAR recommendations are not mandatory, but they are highly recommended and are followed by most operators and all hospitality accounting firms.

To illustrate the production of an income statement using the USAR recommendations, consider Figure 3.2. It shows the USAR suggested income statement format used by Joshua's Restaurant for the year ending December 31, 20xx. Note: The "Line" column has been added by the authors for the reader's ease in identifying data locations.

Joshua's Restaurant

Income Statement

For the Year Ended December 31, 20xx

Line		
1	**SALES**	
2	Food	$ 1,891,011
3	Beverage	$ 415,099
4	**Total Sales**	**$2,306,110**
5	**COST OF SALES**	
6	Food	$ 712,587
7	Beverage	$ 94,550
8	**Total Cost of Sales**	**$ 807,137**
9	**LABOR**	
10	Management	$ 128,219
11	Staff	$ 512,880
12	Employee Benefits	$ 99,163
13	**Total Labor**	**$ 740,262**
14	**PRIME COST**	**$1,547,399**
15	**OTHER CONTROLLABLE EXPENSES**	
16	Direct Operating Expenses	$ 122,224
17	Music and Entertainment	$ 2,306

Figure 3.2 Sample USAR Income Statement

18	Marketing	$ 43,816
19	Utilities	$ 73,796
20	General and Administrative Expenses	$ 66,877
21	Repairs and Maintenance	$ 34,592
22	**Total Other Controllable Expenses**	**$343,611**
23	**CONTROLLABLE INCOME**	**$415,100**
24	**NON-CONTROLLABLE EXPENSES**	
25	Occupancy Costs	$ 120,000
26	Equipment Leases	$ —
27	Depreciation and Amortization	$ 41,510
28	**Total Non-Controllable Expenses**	**$161,510**
29	**RESTAURANT OPERATING INCOME**	**$253,590**
30	Interest Expense	$ 86,750
31	**INCOME BEFORE INCOME TAXES**	**$166,840**

Figure 3.2 *(Continued)*

The USAR income statement is arranged on the above income statement *from the expenses that are most controllable to least controllable* by a foodservice operator. The format of a USAR income statement can best be understood by dividing it into its main sections:

1) Sales (Revenue)
2) Total Cost of Sales
3) Total Labor
4) Prime Cost
5) Other Controllable Expenses
6) Non-Controllable Expenses
7) Income (Profits)

Sales (Revenue)

After listing the name of an operation and the accounting period addressed, the USAR format for an income statement lists sales (Line 1), or revenue, first and on the top line. **Topline revenue** represents the total sales generated by a business prior to any expense deductions.

Historically, food sales (Line 2) include income from the sale of all food items and from the sale of non-alcoholic beverages such as soft drinks, coffee, tea, milk, bottled water, and fruit juices. When alcoholic beverages are sold, the revenue from the sales of these beverage products (Line 3) is added to food sales to yield an operation's **total sales** (Line 4).

When using the USAR, the physical layout of the income statement for different types of business can vary somewhat in their formats. For example, a foodservice operation that serves alcohol would want its income statement to identify the revenue generation and costs associated with serving alcoholic drinks. In a family-style pancake restaurant that serves only food and non-alcoholic beverages, the income statement would not include a line for alcoholic beverages.

The USAR's suggested income statement format recommends, at minimum, that food and (alcoholic) beverage sales be reported separately. There are a variety of reasons for this recommendation. These include the ability to better control costs when these sales are separated, and the requirement of most states to record alcohol sales separately from food sales.

In Figure 3.2, Joshua's food sales are separated from alcoholic beverage sales. In other operations, individual revenue categories may be created for catering sales, on-premises banquet sales, take-out versus dine-in sales, merchandise (e.g., logoed hats, cups, and T-shirts), or any other revenue category that would be helpful to the operation's managers as they analyze their sales.

In some foodservice operations, the sale of logoed and other merchandise types can be significant. Readers familiar with the Cracker Barrel restaurant group may be aware of the significant amount of merchandise it sells. Note: Even if an operation's merchandise sales are relatively small, they should be reported separately on the income statement.

Total Cost of Sales

Food and beverage **cost of sales** (Line 5) are entered separately on the income statement. The reporting of food sales and the related cost of food sales (Line 6) separately from the cost of beverage sales (Line 7) is useful for analyzing a

Key Term

Topline revenue: Sales or revenue shown on the top of the income statement of a business.

Key Term

Total sales: The sum of food sales and alcoholic beverage sales generated in a foodservice operation.

Key Term

Cost of sales: The total cost of the products used to make the menu items sold by a foodservice operation.

foodservice operation. Without this separation an inefficient food operation could be covered up by a highly profitable beverage operation (or the reverse could occur).

The cost of sales reported on the income statement should be their actual cost after considering inventory on hand at the beginning and end of the accounting period, employee meals, and other factors.

In many retail businesses, cost of sales is calculated as:

> Beginning inventory
> + Purchases
> − Ending inventory.
> ──────────────────
> Total Cost of sales

This simple formula is not acceptable for most foodservice operations because the expense of any employee meals provided would be classified as part of the total cost of sales. These meal costs are better accounted for as a labor cost. In addition, in those operations that separate their costs into food costs and beverage costs, the value of food products transferred from the kitchen to the bar (e.g., fruits, juices, and garnishes) must be accounted for, as well as the value of products transferred from the bar to the kitchen (e.g., wine and beers used for cooking).

A better model to calculate cost of sales for food in many foodservice operations is:

Beginning inventory (food)   _____

Plus:

Food purchased during the accounting period   _____

Equals:

Food available for sale   _____

Less:

Ending inventory (food)   _____

Equals:

Cost of food consumed   _____

Less:

Employee meals   _____

Transfers to beverage cost   _____

Plus:

Transfers from beverage cost   _____

Equals:

Cost of sales (food)   _____

The same formula is used to calculate cost of sales for beverage, with the exception that there will not be a reduction for employee meals. A separate income statement listing for "Cost of sales: Food" and "Cost of sales: Beverages" equal to total cost of sales (Line 8) is integral to product control procedures. This topic will be discussed fully in Chapter 8 (Food and Beverage Cost Control).

Find Out More

The USAR contains suggestions for preparing income statements primarily for businesses operated in the United States. However, income statements prepared in other countries can vary in format from the income statements prepared according to USAR recommendations. One good example of this relates to the concept of profits. In the United Kingdom, for example, income statements are routinely prepared in a way that indicates an operation's "gross profit."
Gross profit is calculated as follows:

Total sales − Total cost of sales = Gross profit

Many operators feel gross profit is a key number because it assesses food and beverage sales and the food- and beverage-related product costs that can and should be directly controlled by the manager on a daily basis. Earlier editions of the USAR included gross profit calculations, but the most recent edition does not do so. In many other countries, however, gross profit remains a key entry on the income statement.

The proper preparation of income statements and understanding the concept of gross profit are both good examples of the expanded knowledge many international foodservice operators must have. This is especially so when they are responsible for businesses operating in countries that establish their own preferred accounting systems used for legally reporting a foodservice unit's operating results.

To learn more about the use of gross profit as a key business metric, enter "importance of gross profit in foodservice operations" and review the results.

Total Labor

While **payroll** generally refers to salaries and wages a foodservice operation pays to its employees, the USAR income statement provides greater labor cost-related details.

"Payroll" as shown in Figure 3.2, "Labor" (Line 9), on a USAR-formatted income statement, is separated into three categories:

✓ Management
✓ Staff
✓ Employee Benefits

Key Term

Payroll: The term commonly used to indicate the amount spent for labor in a foodservice operation. Used for example in "Last month our total payroll was $28,000."

Management (Line 10) includes the total amount of salaries paid during the accounting period, and staff (Line 11) refers to payments made to hourly (non-salaried) workers.

The employee benefits category (Line 12) includes the cost of all benefits payments made for managers and hourly workers. Some benefit payments are mandatory [such as FICA (Social Security)], while others are voluntary (such as the cost of providing health insurance).

The specific employee benefits paid by an operation will vary but can include:

✓ FICA (Social Security) taxes, including taxes due on employees' tip income
✓ FUTA (Federal unemployment taxes)
✓ State unemployment taxes
✓ Workers' compensation
✓ Group life insurance
✓ Health insurance, including:
  • Medical
  • Dental
  • Vision
  • Hearing
  • Disability
✓ Pension/retirement plan payments
✓ Employee meals
✓ Employee training expenses
✓ Employee transportation costs
✓ Employee uniforms, housing, and other benefits
✓ Vacation/sick leave/personal days
✓ Tuition reimbursement programs
✓ Employee incentives and bonuses

Not every operation will incur all the benefit costs listed above, but some operations may have all of these and perhaps additional benefits.

Total labor expense (Line 13) is listed prominently on a USAR income statement because controlling and evaluating total labor cost is important in every foodservice operation. In fact, many operators feel it is even more important to control labor costs than product costs. One reason is that labor and labor-related benefit costs comprise a larger portion of their operating costs than do food and beverage product costs.

Key Term

Total labor: The cost of the management, staff, and employee benefits expense required to operate a business.

Technology at Work

In many foodservice operations, the cost of labor is equal to, or even exceeds, the operation's cost of sales. In a foodservice operation, the need for hourly paid employees can vary tremendously at different times of the day and on different days of the week. Therefore, developing and maintaining cost-effective employee schedules is a critical management task.

The scheduling of hourly employees can be time consuming and challenging as operators plan which employees should work and when they should work. This is especially so when the operation employs many part-time workers with varying scheduling needs.

Fortunately, some companies have designed software programs that assist operators in creating and quickly modifying hourly worker schedules. The best of these programs allows employees to access the schedule from their own smart devices so there is no need to return to the operation to learn if staff have been added to or reduced from the work schedule.

Employee shift scheduling software saves time and ensures employees are always scheduled according to the operation's needs and the employees' personal preferences and availability. Basic employee scheduling programs are often offered at no cost, and those with more advanced features are available for lease.

To review the features of free-to-use employee scheduling tools, enter "best free scheduling software for restaurants" in your favorite search engine and review the results.

Prime Cost

Prime cost (Line 14) is defined as an operation's total cost of sales added to its total labor expense. Note: As previously defined, "total cost of sales" represents the amount paid for the food and beverage products sold by an operation, while "total labor" represents the cost of the management, staff, and employee benefits expense required to operate the business.

Key Term

Prime cost: An operation's cost of sales plus its total labor costs.

Prime cost is clearly listed on the income statement because it is an excellent indicator of an operator's ability to control product costs (cost of sales) and labor costs, the two largest expenses in most foodservice operations. The prime cost concept is also important because, when prime costs are excessively high, it is difficult to generate a sufficient level of profit in a foodservice operation even when other controllable and non-controllable expenses are maintained.

While each foodservice operation is different, prime costs in the range of 60–70% of revenue are common for **full-service restaurant** operations, while prime costs below 60% are most common in **fast casual** and **quick-service restaurants** (QSR).

Other Controllable Expenses

As shown in Figure 3.2, after prime cost, the next major section is reserved for identifying other controllable expenses (Line 15): the non-food/non-alcoholic and labor costs controlled by managers incurred from operating the business. There are two major issues that concern managerial accountants as they consider expense data included on a USAR income statement. These relate to the **timing** and the **classification** (placement) of the expense.

Just as a foodservice operator must record all revenue generated during the accounting period addressed by the income statement, so too must the operator make sure all of the related expenses incurred have been reported. If not done properly, expenses would be understated, and profits would be overstated. Alternatively, to intentionally (or unintentionally) overstate expenses and include expenses not incurred would have the effect of understating profits.

Recall from Chapter 1 that accrual accounting requires a business's revenue to be reported when earned and its expenses to be recorded when incurred. This matching principle is designed to closely tie expenses of a business to the actual revenues generated by the expenses. Matching expenses to revenue may, at first examination, seem to be a simple process. However, managerial accountants must often make important decisions to determine the appropriateness of specific expenses on an income statement for a specific accounting period.

Decisions made about how best to match revenues to expenses can be complex and open to

Key Term

Full-service restaurant: A foodservice operation at which servers deliver food and drink offered from a printed menu to guests seated at tables or booths.

Key Term

Fast casual (restaurant): A sit-down foodservice operation with no wait staff or table service. Customers typically order off a menu board and seat themselves or take the purchased food elsewhere.

Key Term

Quick-service restaurant (QSR): Foodservice operations that typically have limited menus and often include a counter at which customers can order and pick up their food. Most quick-service restaurants also have one or more drive-through lanes that allow customers to purchase menu items without leaving their vehicles.

Key Term

Timing (expense): Determining when to place an expense on an income statement.

Key Term

Classification (expense): Determining where to place an expense on an income statement.

honest differences of opinion among managerial accountants. In some cases, answers to questions such as these are established by GAAP (see Chapter 1) or, in other cases, the best decision made by the person(s) preparing the income statement.

To illustrate, assume a foodservice operator produces a monthly income statement. The operator is now preparing an income statement for January, the first month of the operation's fiscal year. The operator pays a property tax bill twice per year. One half of the bill is due in February, and the remaining balance is due six months later in August. For the January income statement being produced, should the operator enter:

✓ 1/12 of the annual bill?
✓ 31/365 of the annual bill?
✓ "$0.00," because no property tax was paid in the month of January?

In this example, the operation also spent $10,000 on produce purchased from a local vendor in December. The operator then received an invoice credit (refund) in January for $200 for some food items of inferior quality delivered in late December of the previous year. The $200 of original product cost was included in the cost of food purchased in December. For January, should the operation's produce expense be recorded as:

✓ $10,000?
✓ $10,000 − $200 = $9,800?

As a final example, assume the operation pays its employees every two weeks. On January 17, it paid its employees for the time worked from December 29 through January 11. Since the entire two-week period was paid in January, should the operator count the entire cost of the payroll as a January expense, or only 11/14th of the payroll (to reflect the 11 actual days in January for which the workers were paid)?

The operator's answers to the above questions directly impact the information presented on their income statements. Regardless of the timing decisions made, recall that the consistency principle (see Chapter 1) of accounting requires managers to be uniform in decision-making. That is, if an expense is treated in a specific manner in one instance, it should be treated in an identical manner in all subsequent situations. Additional factors that may influence the timing of expenses include GAAP, company policies, and rules and regulations enforced by taxing authorities or other governmental agencies.

Like the timing of expenses, the classification of expenses can be simple or complex. For example, if the cost of repairing a broken refrigerator in a foodservice operation is $500 during a specific month, this expense can easily be recorded in the operation's "Repair and Maintenance" expense category. In other cases, expense classification can be more complex. To illustrate, assume the following business transaction:

✓ A foodservice operation's customer places a $100 food delivery order and pays by credit card.
✓ The operation pays a 10% ($10.00) delivery fee to a third-party delivery company.
✓ The operation pays a 20% ($20.00) marketing fee to the third-party delivery company.
✓ The operation pays a 3% ($3.00) credit card fee.
✓ The operation collects sales tax of 8% ($8.00) on the customer's order.

The credit and debit (see Chapter 2) entries to properly record the revenue and expense of the business transaction are shown in Figure 3.3.

As all expense timing and classifications are made, the amounts of these **Other Controllable Expenses** are listed on the income statement. In general, these are expenses that can be influenced by a foodservice operator's own decisions. The USAR identifies and allows the use of several "other controllable expense" categories, and these are identified in Figure 3.2.

The specific other controllable expense categories of an operation listed on its income statement may vary. As seen in Figure 3.2, commonly reported controllable expenses include:

Key Term

Other Controllable Expenses: Expenses that a foodservice operator can influence with increases or decreases based on business decisions. Examples include marketing costs and utility costs.

Direct Operating Expenses (Line 16): In this expense category, operators list the cost of uniforms, laundry, supplies, menus, kitchen tools, and other items incidental to service in the dining areas that provide support in the kitchen and storage areas.

Revenue and Expense Accounts	Debit	Credit
Accounts receivable from third-party	$75.00	
Delivery expense	$10.00	
Marketing expense	$20.00	
Credit card fee expense	$ 3.00	
Food sales		$100.00
Sales tax payable		$ 8.00

Figure 3.3 $100 Sample Food Sale: Journal Entries

Music and Entertainment (Line 17): These costs, if significant, should be shown separately. Many foodservice operations offer little or no music or entertainment. When this is the case and the expenses are small, they may be recorded as "Miscellaneous" within the direct operating expense category.

Marketing (Line 18): This expense category includes newspaper, magazine, radio and TV, and Internet advertising, as well as other expenses for outdoor signs and direct mailings. Loyalty program costs, donations, and special events that promote the business can be additional marketing costs. In some cases, marketing fees associated with third-party meal delivery companies such as DoorDash and Uber Eats are also included in this category.

Utilities (Line 19): This category includes the cost of water, sewage, electricity, and gas used for heating a building. When facilities are rented and the restaurant pays the utility bills, these costs are recorded as "utilities" rather than rent.

General and Administrative Expenses (Line 20): This group of costs includes expenses generally classified as operating "overhead." These expenses are necessary for the operation of the business as opposed to being directly connected with serving guests. This category most often includes the cost of items such as office supplies, postage, credit card fees, telephone charges, data processing, general insurance, professional fees, and security services.

Repairs and Maintenance (Line 21): These expenses include painting and decorating costs, maintenance contracts for elevators and machines, and repairs to an operation's various equipment and mechanical systems. It is not used to record the purchase of new equipment.

The individual entries in Other Controllable Expenses are summed to yield the Total Other Controllable Expenses amount (Line 22). That amount and prime cost are then subtracted from total sales to yield the operation's **Controllable Income** (Line 23).

Controllable income (and its percentage of total sales) is often used to evaluate the effectiveness of unit-level managers because it represents sales minus only those costs and expenses that unit managers generally can directly control or influence. In many foodservice companies, controllable income is the basis for determining at least some portion of a foodservice manager's incentive or bonus pay.

Key Term

Controllable Income: The amount of money remaining after an operation's controllable expenses have been subtracted from its total sales.

Technology at Work

Foodservice operations are very energy-intensive businesses. The average restaurant uses 5–10 times more energy than the average commercial building. This is not surprising when considering the large number of energy-intensive pieces of equipment often used in a foodservice operation.

For example, an operation's back-of-house (kitchen) utilizes equipment such as ovens, stoves, refrigerators, freezers, and mechanical dishwashers. In the front-of-house (public areas), exterior and interior lighting and heating, ventilating, and air conditioning (HVAC) systems consume significant amounts of energy.

Between the ongoing labor shortage, supply chain disruptions, and the shift to "off-site delivery" service models, foodservice operators are adapting to do more with less. Nowhere is this more evident than in their efforts to reduce utility usage and cost.

New technology provides many solutions to optimize power consumption and reduce operating costs. These technologies are known as Energy Management Systems (EMS).

EMS and their providers promote energy efficiency by combining hardware, software, consulting, and control systems to reduce an operation's overall energy use and cost. To learn more about EMS, enter "energy management systems for restaurants" in your favorite search engine and review the results.

Non-Controllable Expenses

Non-controllable expenses (Line 24) are listed next on the income statement in Figure 3.2, and they include all costs not under the immediate control of management.

Non-controllable costs are often referred to as **fixed costs**. Unlike **variable costs**, (those that vary with the amount of sales, such as staff wages), fixed costs remain unchanged regardless of an operation's sales volume.

Non-controllable expenses include those previously committed and must be incurred irrespective of an operation's sales volume. In a foodservice operation, examples of non-controllable expenses include rental payments, interest on long-term loans,

Key Term

Non-controllable expenses: Costs which, in the short run, cannot be avoided or altered by management decisions. Examples include lease payments and depreciation.

Key Term

Fixed cost: An expense that stays constant despite increases or decreases in sales volume.

Key Term

Variable cost: An expense that generally increases as sales volume increases and decreases as sales volume decreases.

depreciation (see Chapter 2), and **amortization** charges.

Non-controllable expenses listed in Figure 3.2 include:

Occupancy costs (Line 25): These include the cost of renting buildings and land, property taxes, and insurance on **fixed assets**. These expenses can vary considerably between foodservice operations. Since the owners of a business most often control these costs, they are only infrequently under the direct control of those operating the business.

Equipment leases (Line 26): These expenses include the cost incurred from the leasing or renting of equipment used in a foodservice operation. Common examples include charges for the leasing of POS systems, beverage dispensing equipment, and ice machines.

Depreciation and amortization (Line 27): These **non-cash expenses** are the result of depreciating buildings and **furniture, fixtures, and equipment (FF&E)**. In addition, this expense includes the amortization of leaseholds and leasehold improvements. Note: Owners/corporate board members usually decide the proper amount of these expenses.

To illustrate the profit variability caused by non-controllable expenses, assume two foodservice operators manage virtually identical units. Both units generate $1 million in sales and have $250,000 in controllable income. One operation has total non-controllable expense of $100,000 and generates a **restaurant operating income** of $150,000 ($250,000 controllable income − $100,000 non-controllable expenses = $150,000 restaurant operating income).

Key Term

Amortization: The practice of spreading an intangible asset's cost over that asset's useful life.

Key Term

Occupancy costs: Costs related to occupying a space including rent, real estate taxes, personal property taxes, and insurance on a building and its contents.

Key Term

Fixed asset: An asset such as land, building, furniture, and equipment that is purchased for long-term use and is not likely to be converted quickly into cash.

Key Term

Non-cash expense: Expenses recorded on the income statement that do not involve an actual cash transaction. Examples include depreciation and amortization, which are expenses where an income statement charge reduces operating income without a cash payment.

Key Term

Furniture, fixtures, and equipment (FF&E): Movable furniture, fixtures, or other equipment that have no permanent connection to a building's structure.

Key Term

Restaurant operating income: All an operation's revenue minus its controllable and non-controllable expenses.

The second operation has total non-controllable expense of $200,000 and generates a restaurant operating income of $50,000 ($250,000 controllable income − $200,000 non-controllable expenses = $50,000 restaurant operating income). In this example, it is likely that both operators are doing good jobs managing their businesses even though the restaurant operating income of the first unit is three times that of the second unit.

The individual entries in the Non-Controllable Expenses section of the income statement are summed to yield Total Non-Controllable Expenses (Line 28). That amount is then subtracted from Controllable Income to yield the operation's Restaurant Operating Income (Line 29).

Profits (Income)

It is interesting that the word "profit" does not actually appear anywhere on a USAR income statement. Some foodservice operators consider "Restaurant Operating Income" (Line 29) to be their business's profit because it represents an operation's sales (revenue) minus all controllable and non-controllable expenses, and it reflects the profit formula introduced earlier in this chapter:

Revenue − Expense = Profit

Restaurant Operating Income in Line 29 does not consider any **interest expense** payments made by a business. Interest Expense (Line 30) is the cost of borrowing money and is recorded on this line of the income statement even if the interest incurred has not been paid, and unpaid interest is accrued at the end of the applicable accounting period.

Restaurant operating income represents all an operation's revenue minus its controllable and non-controllable expenses, and it may be further adjusted to account for corporate overhead, interest expense, or other owner-controlled expenses.

Income Before Income Taxes (Line 31) on the income statement is calculated as Restaurant Operating Income minus Interest Expense.

When an operation's revenue exceeds its expenses, the amount of the income before income taxes is

Key Term

Interest expense: The cost of borrowing money.

Key Term

Income Before Income Taxes: The amount of money remaining after an operation's interest expense is subtracted from the amount of its restaurant operating income. Also, a business's profit before paying any income taxes due on the profits.

shown on the income statement. However, if expenses *exceed* revenue, the resulting negative numbers (losses) on the income statement are typically designated in one of the three ways:

1) By a minus ("−") sign in front of the number. For example, a $1,000 loss would be presented on the statement as −$1,000.

2) By brackets "( )" around the number. For example, a $1,000 loss would be presented on the statement as ($1,000).

3) With red ink rather than black ink to designate the loss amount. For example, a $1,000 loss would be presented on the statement as "$1,000," but the number would be printed in red. Note: This approach gives rise to the slang phrase to "operate in the red:" a business that is not making a profit. In a similar vein, to "operate in the black" indicates the business is profitable.

In Figure 3.2, the operation's Income Before Income Taxes (Line 31) is a number that is frequently referred to as a business's "profit." Note that the sample income statement shown in Figure 3.2 does *not* include a line for income taxes due. If the business was operating as a corporation, which is a taxable entity, then an additional expense line for income taxes would be included in the income statement.

Many foodservice operations, however, are operated as sub S corporations, limited liability companies, partnerships, or sole proprietorships. In these entities, taxable income or losses flow through the business and are included on the income tax returns of the owners, shareholders, members, and partners. For this reason, the income statements prepared for these non-taxable entities reflect income up to, but not including income taxes due.

EBITDA

One key financial metric that is *not* listed on a USAR income statement is **EBITDA** (earnings before interest, taxes, depreciation, and amortization).

Most foodservice owners are interested in knowing their operations' ability to generate cash. An operation's EBITDA is a measure that business owners can use to determine their net cash (before taxes) income.

EBITDA is a unique number because it does not include the non-cash operating expenses of interest, taxes, depreciation, and amortization. Payments for these items are listed on the income statement, or elsewhere on a foodservice operation's financial statements, but they are not related to the day-to-day core operation of a business. For example, the interest paid on debts is listed on the income statement, but this expense reflects how the business was financed, and not the ability of the business to generate sales or profits. Similarly, income taxes due may be, or may not be,

Key Term

EBITDA: Short for "earnings before interest, taxes, depreciation, and amortization." EBITDA is used to track and compare the underlying profitability of a business regardless of its depreciation assumptions or financing choices.

listed on an operation's income statement, but these do not affect how managers operate on a daily basis.

Calculating EBITDA from a USAR formatted income statement is an easy two-step process.

Step 1: Identify Restaurant Operating Income

In Figure 3.2, Restaurant Operating Income is listed as $253,590 (see Line 29).

Step 2: To Line 29, add back the amount listed for Depreciation and Amortization (see Line 27):

Restaurant Operating Income	$253,590
+ Depreciation and Amortization	$ 41,510
= EBITDA	$295,100

In this example, EBITDA for Joshua's Restaurant = $295,100.

Many foodservice operators view EBITDA as a good way to assess the earning power of their businesses and to compare their own operations with similar operations having different debt levels or depreciation amounts.

However, it is also important to recognize important limitations of EBITDA. For example, interest and income taxes are actual business expenses that must be paid. In addition, depreciation and amortization should reflect the decline in the real value of a business's assets. Each of these may be "non-cash" expenses, but the expenses are real! In general, most owners agree that EBITDA is one good way to help assess the cash generating ability of a restaurant.

What Would You Do? 3.1

"So which is it?" asked Lani, the dining room manager at the 200-seat Harvest House Restaurant, "were our sales up or down last month compared to the previous year?"

"Both" replied Jackson, the manager of Harvest House.

"How can they be up and down at the same time?" asked Lani.

"Well," replied Jackson, "our overall topline sales last month were about the same as the prior year, but when I looked into it closer, and checked the POS system, our actual dine-in sales were lower, and our carry-out sales were higher."

Assume you were the owner of Harvest House. How important would it be to you to know the proportion of your operation's sales that were coming from dine-in versus carry-out business? Would you recommend that the USAR income statement prepared for your operation be modified to show these two food revenue categories separately? Explain your answer.

Utilizing the Income Statement

To properly read and analyze an income statement, foodservice managers must recognize that the USAR income statement shown in Figure 3.2 is an **aggregate statement**. This means that all details associated with the sales, costs, and profits of the foodservice establishment are summarized on the statement.

Although an aggregate statement provides a foodservice operator with an overall look at the performance of an operation, detailed expenses within each cost category are not included directly on the statement. These details can be found in the income statement's **supporting schedules.**

Key Term

Aggregate statement: A financial statement in which data from many individual financial accounts are summarized in one document.

Key Term

Supporting schedule: A detailed itemization of the contents of an account.

Supporting Schedules

The USAR, the Uniform System of Accounts for the Lodging Industry (USALI), and the Uniform System of Financial Reporting for Clubs (USFRC) all recommend the use of supporting schedules. The methods managerial accountants in these industries use to create appropriate schedules are important areas addressed by these uniform systems. It is essential to recognize, however, that managers should always use the specific methods of reporting that best maximize and clarify the information provided to the statement's readers.

Information about how much revenue is generated and the associated costs incurred to do so are critical to many statement readers, and this information should be presented clearly, honestly, and in keeping with GAAP (see Chapter 1). To do otherwise is dishonest and unethical, and it may also be illegal!

Each important line item on the income statement should be accompanied by a supporting schedule that outlines what an operator must know about that category to operate the business successfully. For example, in Figure 3.2, Direct Operating Expenses (Line 16) could have a supporting schedule detailing costs incurred for uniforms, laundry and linen, china and glassware, silverware, and the like.

Figure 3.4 shows a supporting schedule that might accompany Joshua's Restaurant's income statement shown in Figure 3.2. Note that the total Direct Operating Expenses of $122,224 in Line 16 of Figure 3.2 matches exactly with the total expenses (bottom line) in the Direct Operating Expense Schedule (Figure 3.4) that accompanies the Joshua's Restaurant's income statement.

Joshua's Restaurant: For the Year Ended December 31, 20xx

Type of expense	Expense	% of Total	Operator's Notes
Uniforms	$13,407	11.0%	
Laundry and linen	40,964	33.5%	
China and glassware	12,475	10.2%	Expense is higher than budgeted because china shelf collapsed on March 22.
Silverware	2,944	2.4%	
Kitchen utensils	7,250	5.9%	
Contract cleaning	2,542	2.1%	
Cleaning supplies	8,571	7.0%	
Paper supplies	2,675	2.2%	
Bar uniforms	5,413	4.4%	
Menus and wine lists	11,560	9.5%	Expense is lower than budgeted because the new wine supplier agreed to print the wine lists without charge.
Exterminating	1,803	1.5%	
Flowers and decorations	9,015	7.4%	
Licenses	3,605	2.9%	
Total	**$122,224**	**100%**	

Figure 3.4 Direct Operating Expenses Schedule

Foodservice operators should create schedules addressing relevant sales and cost areas when they assist in management decision-making and enhance and clarify the reader's understanding of the income statement.

Even when using USAR recommendations to create an income statement, the actual format of the statement can vary depending on the unique needs of a specific foodservice operation. In all cases, however, the creation of a properly prepared income statement should adhere to the principles of income statement preparation shown in Figure 3.5.

Comparative Analysis

An income statement summarizes a foodservice operation's performance for a specific time period. While a properly prepared income statement provides an accurate summary of how an operation performed, it makes no declaration about the quality of the operation's performance.

A properly prepared income statement:

1) Clearly identifies the business whose revenues and expenses are being summarized.
2) Plainly states the specific accounting period for which the statement is prepared.
3) Includes a summary in the most informative (detailed) manner practical, of all revenue generated by the business during the accounting period.
4) Summarizes all accounting period expenses utilized by the business to generate the stated revenue.
5) Utilizes a logical and consistent system to classify expenses.
6) Provides additional clarity using supporting schedules where appropriate.
7) Incorporates the use of an appropriate uniform system of accounts.
8) Applies GAAP in its preparation.

Figure 3.5 Principles of Income Statement Preparation

Those who read an income statement typically do so to learn the profitability of a business relative to operational efficiency and sales levels and to assess whether costs are appropriate when compared to the amount of revenue generated.

Income statement readers might determine the quality of an operation's performance related to four different sources of information:

✓ Similar units
✓ Industry standards
✓ Previous performance
✓ Budgeted performance

Note: Readers of income statements may (or may not) have access to operating data related to similar units and, perhaps even, to industry standards. However, information regarding comparisons of a foodservice operation's previous and budgeted performance will likely be available to those with access to the operation's income statement.

Similar Units

One good way to assess a foodservice operation's performance is to compare its income statement results to those of similar units. However, in most cases, a foodservice operator will not share their operating results with their competitors. For those foodservice professionals managing a **franchised operation**, however, the franchisor will often, if possible, provide a summary income statement including financial results of similar franchised units operating in a nearby area.

Key Term

Franchised operation:
A method of distributing products or services involving a franchisor, who establishes the brand's trademark or trade name and a business system, and a franchisee, who pays a royalty and often an initial fee for the right to operate under the franchisor's name and system.

Informally, operators may also share some sensitive financial information with others who are members of a local or a state restaurant association.

It is important to note that every foodservice operation is unique. However, financial information from similar-sized facilities with the same service styles can often help operators assess their own profit-making performance.

Industry Standards

In some cases, a foodservice operator may compare operating results to industry standards (averages) published on a regular basis. Published standards often include averages for percentages of revenue from alternative revenue sources such as on-premises dining and carry-out. In addition, cost of sales for food and beverages are often reported as industry averages. Foodservice operators searching online may also find statistics related to average levels of controllable costs (such as labor) and fixed costs (such as occupancy costs) for a specific style of operation.

While compiled and published industry standards may reflect accepted ranges of average sales and expense, they are not exact measurements and must be carefully evaluated to determine their applicability to a specific foodservice operation.

Technology at Work

Several non-profit and for-profit entities compile statistics on the foodservice industry. For example, the National Restaurant Association (NRA) is a non-profit association of restaurant operators, and it publishes an annual report summarizing performance of the entire restaurant industry. The NRA also provides forecasts of future revenue growth, and much of this data is at no cost to NRA members.

Other for-profit entities compile foodservice industry statistics that can be useful to operators. For example, Technomics is one of the oldest and largest organizations that tracks sales levels for industry-leading chains and brands that include data from over 90% of the franchised foodservice industry. It also monitors menu trends of thousands of menus on an interactive online interface, and this data can help operators identify flavors, ingredients, and menu items that are growing or shrinking in popularity nationwide.

To review some of the key industry data compiled by Technomics, enter "Technomics foodservice statistics" in your favorite search engine and review the results.

Previous Performance

One of the best ways for a foodservice operator to utilize results from the income statement is to compare financial information to those of the same operation in a prior accounting period. For example, foodservice operators reviewing their income statement for October of this year can compare the results with the results from October of the previous year. Similarly, an operator preparing a year-end income statement can compare their operating results to those of the prior year.

When making comparisons to previous performance, operators assess increases or decreases in sales, controllable expenses, and profits to determine where improvements have been made and where management's attention must be directed to correct any current deficiencies in the operation's performance.

It is important to note that an operator must consistently apply GAAP in the production of each income statement to ensure comparisons are based on the same variables.

Budgeted Performance

For most foodservice operators, the best way to analyze an income statement is to compare it to previously planned or budgeted performance. For example, when an operating budget has been prepared, foodservice operators can compare their actual operating results against the budgeted (planned) results. Note: The creation and management of operating budgets is addressed in this book in Chapter 11 (Operating Budgets).

When reviewing income statements compared to budgeted results, operators can begin to answer many important operational questions such as:

1) How did our actual sales compare to our budgeted sales?
2) Were our actual cost of sales for food and beverages higher or lower than budgeted cost of sales?
3) Were actual labor costs in line with budgeted labor costs?
4) Did our total controllable costs exceed our budgeted controllable costs?
5) Were our profits in line with our profit forecasts?

To compare an operation's actual performance with its budgeted performance requires the preparation of an accurate and thoughtfully produced budget. When such a budget exists, operators can gain tremendous insight as they assess variations between actual results and planned results. The creation of a foodservice budget that will permit perceptive comparisons to actual operating results is so essential that it will be the sole topic of Chapter 8 in this book.

The income statement is one of three key financial statements required to be produced when a business utilizes GAAP. The balance sheet (see Chapter 2) is another of the required financial statements and its production and use are the topics of the next chapter.

What Would You Do? 3.2

"I don't understand what you're so concerned about," said Charles Lohr, the bar manager at the Chicken Zone.

The Chicken Zone was a sports bar that featured a wide variety of flavored chicken wings and other fried chicken products served in a casual setting. The Chicken Zone catered to sports fans coming to watch games on one of the operation's 20 large video screens. Beer and mixed drinks were popular accompaniments to the operation's food menu.

Charles was talking to Andrea, the restaurant's General Manager, about the previous month's performance as shown on its recently produced monthly income statement.

"Our projected beverage sales for last month was $80,000, and we hit that target almost exactly," said Charles.

"Yes, we hit our sales target," replied Andrea, "and I'm pleased with that. It's our cost of beverage that has me concerned."

"Well, we projected a beverage cost of sales of $20,000, and as I read the income statement, we only spent $13,000 on beverages; so that should be really good news!" said Charles.

Assume you were Andrea. What could be possible explanations for why her operation's cost of beverage sales would be so low compared to its originally forecasted amount? Do you agree with Charles that the reduced cost of beverage sales is actually "really good news"? Explain your answer.

Key Terms

Profit and loss (P&L) statement

Profit center

Business dining

Non-commercial foodservice

Cost center

Stakeholder

Appreciation

Return on investment (ROI)

Creditor

28-day accounting period

Topline revenue

Total sales

Cost of sales

Payroll

Total labor

Prime cost

Full-service restaurant

Fast casual (restaurant)

Quick-service restaurant (QSR)

Timing (expense)

Classification (expense)

Other Controllable Expenses

Controllable Income

Non-controllable expenses

Fixed cost

Variable cost

Amortization

Occupancy costs

Fixed asset

Non-cash expense

Furniture, fixtures, and equipment (FF&E)

Restaurant operating income

interest expense

Income before income taxes

EBITDA

Aggregate statement

Supporting schedule

Franchised operation

Operator's 10-Point Tactics for Success Checklist

Evaluate your need for, and the current status of, each of the following operational tactics. For those tactics you think are important, but not yet in place, develop an action plan for its implementation including who will be responsible for the tactic's completion and the target date by which it should be completed.

Tactic	Don't Agree (Not Done)	Agree (Done)	Agree (Not Done)	If Not Done	
				Who Is Responsible?	Target Completion Date
1) Operator realizes the importance of income statements to owners, investors, lenders, and creditors.	____	____	____		
2) Operator understands the differences between fiscal year, calendar year, monthly, and 28-day accounting period income statements.	____	____	____		
3) Operator recognizes the purpose of reporting sales separately for food items and alcoholic beverage items.	____	____	____		
4) Operator recognizes the purpose of calculating cost of sales separately for food items and alcoholic beverage items.	____	____	____		
5) Operator knows that the three components of "Total labor" as listed on an income statement are Management, Staff, and Employee Benefits.	____	____	____		
6) Operator can identify operating expenses that should be listed under "Other Controllable Expenses" on an income statement.	____	____	____		

Tactic	Don't Agree (Not Done)	Agree (Done)	Agree (Not Done)	If Not Done	
				Who Is Responsible?	Target Completion Date
7) Operator can identify those expenses that should be listed under "Non-Controllable Expenses" on an income statement.	——	——	——		
8) Operator recognizes the purpose of supporting schedules when reading and analyzing an income statement.	——	——	——		
9) Operator understands how to calculate EBITDA from a USAR-formatted income statement.	——	——	——		
10) Operator recognizes income statement performance can be compared to the performance of similar foodservice units, industry standards, prior period operation performance, or budgeted performance.	——	——	——		

4

The Balance Sheet

What You Will Learn

1) The Importance of the Balance Sheet
2) How Assets Are Reported on the Balance Sheet
3) How Liabilities and Owners' Equity Are Reported on the Balance Sheet
4) How to Analyze a Balance Sheet

Operator's Brief

In this chapter, you will learn why the balance sheet is an extremely important financial document. As a foodservice owner or operator, the ability to read and understand a balance sheet is second in importance only to the income statement. You have learned that the income statement provides a summary of a business's operational results over a defined period (e.g., a month or a year); the balance sheet reflects the overall financial condition of a business on the day it was prepared. Therefore. it is a point-in-time "snapshot" of a business's financial standing and value on that date.

In Chapter 2, you also learned that the accounting equation has three distinct components:

$$Assets = Liabilities + Owners'\ Equity.$$

A balance sheet tells its readers as much as possible about these three basic accounting equation components. It also provides readers with details about the precise nature and condition of a business's assets, liabilities, and owners' equity.

In this chapter, you will learn why owners, investors, lenders, and managers must know how to read and understand a balance sheet. You will also learn how accountants prepare a balance sheet using a Uniform System of Accounts for Restaurants (USAR) format. Most importantly, you will learn how managerial accountants evaluate business information using financial ratios and vertical analysis.

CHAPTER OUTLINE

The Importance of the Balance Sheet

Managerial accountants prepare balance sheets to better understand the value of a business and how well its assets were utilized to produce wealth for the business's owners. Also known as the "Statement of Financial Position," the balance sheet reports the assets, liabilities, and net worth of a foodservice operation at the point in time that it was prepared (generally the end of a month or year).

The balance sheet has several important purposes including telling the amount of **cash on hand** at the end of an accounting period. All assets owned by a business should eventually be converted, either directly or indirectly, to cash. Cash on hand is money available for a variety of future uses such as paying bills, servicing debt, and dispersing profits to owners.

Key Term

Cash on hand: The amount of money a business has available to spend on the last day of an accounting period.

The balance sheet also gives details about an operation's fixed assets (see Chapter 3). Further, it reviews the composition of debt and net worth (if a sole proprietorship) or stockholders' equity (if a corporation). This reflects how a foodservice operation has been financed. Generally, the greater the amount of debt financed with equity (see Chapter 2), the greater the financial risk. A foodservice operation using debt to finance a large percentage of its fixed assets may also have a difficult time securing financing for more debt.

The balance sheet also shows the amount of an operation's past earnings retained in the business. For a corporation, retained earnings is the sum of past earnings retained in the business. For a sole proprietorship, past earnings are part of the owner's capital. For example, on a balance sheet, the account for James Smith, the sole owner of a foodservice business, would simply be "James Smith, Capital." Generally, the greater the retention of internally generated funds, the less borrowing will be required during an expansion period.

The balance sheet provides insight into a food-service operation's ability to pay its bills in a timely manner, and **liquidity** is important when timing **cash flows**.

The past several years have seen a renewed interest in the balance sheet. Some start-up businesses that consistently reported profits on their income statements have gone bankrupt a short time later because they were unable to pay their bills when they were due. The balance sheet reveals the amount of an operation's debt, which is critical to understanding the financial strength of a business.

Key Term

Liquidity: A measure of a foodservice operation's ability to convert assets into cash.

Key Term

Cash flows: The net amount of cash and cash equivalents being transferred in and out of a business. Cash received represents inflows, and money that is spent represents outflows.

Users of the Balance Sheet

Several entities are interested in balance sheet information. To better understand why, consider John Calvin. He is the sole owner of a popular foodservice operation that last year grossed $1,000,000 in sales and generated $100,000 in after-tax profit. Louise Schneider, one of his competitors, also grossed $1,000,000 in sales and generated $100,000 in after-tax profit in her operation. Therefore, if the income statements for each operation was compared, it might appear that both businesses were "equal" in value (worth) because each produced a $100,000 after-tax profit.

In fact, however, a closer look at the basic accounting equation (see Chapter 1) for each of the two operations reveals the following:

John's Basic Accounting Equation:

$$\text{John's Assets} = \text{Liabilities} + \text{Owners' Equity}$$

Or

$$\$2,000,000 = \$500,000 + \$1,500,000$$

Louise's Basic Accounting Equation:

$$\text{Louise's Assets} = \text{Liabilities} + \text{Owners' Equity}$$

Or

$$\$2,000,000 = \$1,500,000 + \$500,000$$

Assume one of the owners want to sell their business for $2,000,000. John "owes" less ($500,000) of his operation's value to others than does Louise ($1,500,000). Put a different way, if both owners sold their operations for $2,000,000, John would, after paying all his restaurant's debts, receive $1,500,000 ($2,000,000 assets − $500,000 liabilities = $1,500,000) while Louise would receive only $500,000 ($2,000,000 assets − $1,500,000 liabilities = $500,000) for her business.

Clearly, these two businesses are different in important ways to their owners, and these differences can be identified only by study of their balance sheets. In fact, the type of information contained on a business's balance sheet is of critical importance to a number of different groups including:

✓ Owners
✓ Investors
✓ Lenders
✓ Creditors
✓ Managers/Operators

Owners

The balance sheet prepared at the end of a defined accounting period lets business owners know about the amount of their business they actually "own." To illustrate, assume an operator purchased a foodservice business for $500,000 and borrowed all the money to do so from a local bank.

This new operator technically "owns" the business, and the keys to it are in the operator's possession. As the owner, the operator may set the opening and closing hours of the business as desired, and he or she can offer any menu items for sale they wish to offer. The local bank, however, can also be considered an owner since the new operator could not sell the business to another without first paying off the bank's $500,000 **lien** against the business.

A lien is the legal right to hold another's property to satisfy a debt. In this example, the bank's lien against the business is a liability. All liabilities must be subtracted from the value of a business before its owners can determine the amount of their own equity (free and clear ownership). The balance sheet is designed to show, among other things, the amount of a business owner's free and clear ownership.

Key Term

Lien: A legal right acquired on one's property by a creditor. A lien generally stays in effect until the underlying obligation to the creditor is satisfied. If the underlying obligation is not satisfied, the creditor may be able to take possession of the property involved.

Investors

As explained in the previous chapter, in most cases investors seek to maximize the return of their investments (ROIs). When a business's balance sheet from one accounting period is compared to its balance sheet covering another time period, investors can measure and evaluate their ROIs.

To illustrate, assume a person invested $50,000 in a foodservice operation. Also assume that, exactly one year later, the person's share of the owners' equity based upon the operation's balance sheet is $60,000.

The value of the investment has grown $10,000 during the year ($60,000 − $50,000 = $10,000). Using the ROI formula presented in Chapter 3, the ROI for the year would be computed as:

$$\frac{\text{Money Earned on Funds Invested}}{\text{Funds Invested}} = \text{ROI}$$

In this example:

$$\frac{\$10,000}{\$50,000} = 0.20 \text{ or } 20.0\% \text{ ROI}$$

ROI calculations are important, and investors need information contained in the balance sheet to accurately compute their annual ROIs.

Lenders

Lenders are most concerned about a business's ability to repay its debts. Consider two similar foodservice operations. One is owned by Ezat and the other by Arlene. Both properties have outstanding long-term mortgages totaling $8,000,000. Both owners would like to borrow an additional $1,000,000 from their local bank to refurbish their dining rooms. Both operations currently have $60,000 monthly mortgage payments, and both foodservice operations show an income before income taxes of $120,000 per year on their income statements.

The basic accounting equation as shown on the balance sheet for Ezat's operation is:

$$\text{Assets} = \text{Liabilities} + \text{Owners' Equity}$$

Or

$$\$11,000,000 = (\$8,000,000 \text{ mortgage liability} + \$2,880,000 \text{ other liabilities}) + \$120,000$$

In this example, Ezat has total liabilities of $10,880,000 (an $8,000,000 mortgage plus $2,880,000 other liabilities) and owners' equity of $120,000. Put another way, if, for an extended time period, Ezat's operation did not actually make a profit based on his income statement, and he elected to use his owners' equity account to fund his monthly mortgage payments, he could make two such payments ($120,000 ÷ $60,000 = 2) before that account would be depleted.

The basic accounting equation as shown on the balance sheet for Arlene's operation is:

$$\text{Assets} = \text{Liabilities} + \text{Owners' Equity}$$

Or

$$\$11,000,000 = \$8,000,000\,\text{mortgage} + \$3,000,000$$

In this example, Arlene has total liabilities of only $8,000,000 (her mortgage) and owners' equity of $3,000,000. Put another way, if, for an extended period of time, Arlene's operation did not make a profit, she could use her owners' equity to make up to 50 mortgage payments ($3,000,000 ÷ $60,000 = 50) before her owners' equity account was depleted.

As a result, if bank officials were considering a new $1,000,000 loan for each operator, Arlene's situation would be viewed much more favorably by the bank's loan approval committee than would Ezat's. The reason: The bank would likely feel better about Arlene's ability to withstand a prolonged economic downturn than they would Ezat's ability to continue his payments with the same economic pressures.

Lenders are always interested in a borrower's ability to repay debts. It is for situations such as Arlene's and Ezat's that lenders read the balance sheet of a business carefully to understand the financial strength (repayment ability) of the business.

Creditors

Foodservice operations are often extended credit by a variety of product and service suppliers. Business creditors, much like lenders, are concerned about repayment. For example, when a supplier agrees to sell food to a restaurant for credit, it expects the restaurant will ultimately be able to pay its bill. In this situation, it would not be unreasonable for the supplier to ask to see the restaurant's balance sheet before a decision was made regarding the wisdom of extending credit to it. The balance sheet is the financial document that most accurately indicates the long-term ability of a business to repay a vendor who has extended credit to that business.

Managers/Operators

As indicated in the previous chapter, managers and operators are often more interested in the information on their income statements than on their company's balance sheets. However, there are several important reasons why they must be able to read and analyze balance sheets to determine the current financial balances of cash, accounts receivable (AR), inventories, payable, and other accounts with direct impact on operations.

For example, the balance sheet states the amount of cash on hand at the time it is prepared. Chapter 5 (The Statement of Cash Flows) explains in detail the extremely vital role cash plays in operating a business. It is clear, however, that a business must consistently have sufficient cash on hand to make timely payments to employees, vendors, and the government (taxes).

In addition, as addressed in Chapter 1, AR refers to the amount of money owed to a business by customers. Managers must be able to read a balance sheet to determine the total amount of AR owed to the business on the date the balance sheet is prepared. While in most foodservice operations guests pay for their food and beverages on the day items are purchased, it is not uncommon for some restaurants to extend credit to special customers. When this occurs, an operation's AR may be substantial.

Even if the income statement indicates a business is very profitable, excessively large amounts of AR (which are *not* identified on the income statement) may be a problem because it is a sign that:

✓ Too much credit has been extended
✓ Credit collection efforts may need review and improvement, if necessary
✓ Cash reserves could become insufficient to meet the short-term needs of the business

Limitations of the Balance Sheet

The balance sheet is important to readers because it reveals, at a fixed point in time, the amount of wealth a company possesses. Wealth in this case is simply defined as the current value of all a company's assets minus the company's obligations. While it might seem that establishing the value of a company's assets and liabilities would be a straightforward process, managerial accountants do not always agree about the best way to do it.

To understand why differences of opinion exist, consider a foodservice operation located in a building in the downtown area of a large city. If its owner were required to place a monetary value on this operation, they might consider:

An historical approach: In this situation, the owner would likely value the operation at the price the owner paid for the property when they bought it.

A current (replacement) value approach: In this situation, the owner would likely value the operation at the amount of money required to fully replace it. This approach is sometimes referred to as the "fair market value" (see Chapter 1) approach because it estimates the price at which an asset could be bought or sold in a current transaction between willing parties.

A future value approach: In this situation, the owner would likely value the property at the amount of money (income) it could earn for the company's shareholders in the future.

While each approach above could make sense with specific assets, no single approach to valuing assets is used by accountants in the preparation of the balance sheet. Knowledgeable balance sheet readers recognize that accountants utilize several evaluation approaches, each of which may make the most sense for specific asset types based upon circumstances and available information.

It is also important to note that balance sheets have been criticized because some company assets are not valued. To understand why, consider that, of all the assets listed on the balance sheet, none consider the relative value, or worth, of a foodservice operation's staff or its managers.

While many companies are fond of saying that "Our employees (staff) are our most important asset," the value of experienced, well-trained staff members is not quantified on the balance sheet. To further clarify this, consider Ahmed. He owns his own restaurant and is contemplating spending $1,000 to replace a food mixer in his bakeshop.

His alternative is to spend $1,000 to send his entire five-person food production team to a one-day food safety training course hosted by his local restaurant association. If Ahmed decides to purchase the food mixer, the value of his assets (Property and Equipment) on the balance sheet will increase; if he elects to "invest" the $1,000 in his staff with the training program, no balance sheet increase will occur. However, the training class, not the up-graded mixer, will yield a greater positive impact on addressing the food quality and food safety issues Ahmed faces. For that reason, astute operating managers and accountants should recognize both the value and limitations of balance sheets when they review them.

USAR Balance Sheet Format

Just as there is a USAR-recommended format for the income statement, there is also a USAR-recommended format for the balance sheet. To illustrate the production of a balance sheet using the USAR recommendations, consider Figure 4.1.

It shows the USAR-suggested balance sheet format used by Peggy's Restaurant for the year ending December 31, 20xx. Note: The "Line" column has been added by the authors for ease in data location identification. An operation's actual balance sheet format may vary slightly, but Figure 4.1 is indicative of all USAR-suggested balance sheets.

Assets

Assets (Line 1) on the balance sheet begins to complete the left side of the basic accounting equation:

Assets = Liabilities and Owners' Equity

Assets are normally segmented into four general categories: Current Assets, Property and Equipment, Investments, and Other Assets. Like most small

Peggy's Restaurant

Balance Sheet

As of December 31, 20xx

Line		$	$	$
1	**Assets**			
2	**CURRENT ASSETS**	$	$	$
3	Cash:			
4	House Banks	6,500		
5	Cash in Banks	35,800		
6	Accounts Receivable:		42,300	
7	Credit Cards	13,800		
8	Customer House Accounts	8,500		
9	Inventories:		22,300	
10	Food	18,000		
11	Beverage	15,600		
12	Supplies & Other	7,500	41,100	
13	Prepaid Expenses		18,500	
14	**Total Current Assets**			124,200
15	**PROPERTY AND EQUIPMENT**			
16	Leasehold Improvements		321,400	
17	Furniture, Fixtures, and Equipment		133,000	
18	Less: Accumulated Depreciation and Amortization		−68,500	
19	**Total Property and Equipment—Net**			385,900
20	**OTHER ASSETS**			
21	Cost of Liquor License		15,800	
22	Lease and Utility Deposits		7,500	
23	**Total Other Assets**			23,300
24	**TOTAL ASSETS**			**$533,400**
25	**Liabilities and Shareholders' Equity**			
26	**CURRENT LIABILITIES**			
27	Accounts Payable		38,600	
28	Payroll Taxes Payable		21,900	
29	Sales Taxes Payable		27,800	

Figure 4.1 Sample USAR Balance Sheet

30	Gift Cards Payable	11,500	
31	Accrued Expenses	10,000	
32	Current Portion, Long-Term Debt	27,600	
33	**Total Current Liabilities**		137,400
34	Long-Term Debt, Net of Current Portion		168,000
35	**Total Liabilities**		305,400
36	**SHAREHOLDERS' EQUITY**		
37	Capital Stock	35,000	
38	Retained Earnings	193,000	
39	**Total Equity**		228,000
40	**TOTAL LIABILITIES AND SHAREHOLDERS' EQUITY**		**$533,400**

Figure 4.1 (*Continued*)

foodservice businesses, Peggy's Restaurant has no investments made by the business, other than in the operation itself, so this asset category is not illustrated in Figure 4.1.

Marketable securities and investments that could be listed in the asset segment of a balance sheet are stocks and bonds the business purchases from *other* companies. These are not to be confused with the value of a company's own stocks listed on its balance sheet as Owners' Equity.

Foodservice operators should understand the current assets portion of the balance sheet because current assets and current liabilities are the two balance sheet areas most directly affected by the actions of managers rather than owners.

CURRENT ASSETS (Line 2) are the first asset type shown on the balance sheet in Figure 4.1. Current assets are cash, assets that could readily be converted to cash, and other assets expected to be used by the business within one year. Current assets are normally listed in the order of their liquidity. Cash (Line 3) is listed first followed by the next current asset that can be most quickly converted to cash. The major current asset accounts are cash contained in House Banks (Line 4) and Cash in Banks (Line 5).

Cash in "House Banks" is the currency and coins kept in cashier banks (used for making guests' change), **petty cash** funds, and any undeposited cash receipts.

"Cash in Banks" is the value of all money held in a business's various bank accounts. It is

Key Term

Petty cash: Money kept on hand to purchase minor items when the use of a check or credit card is not practical.

common for different bank accounts to be shown separately on the balance sheet. For example, in addition to a general banking account, a foodservice operation may maintain a separate payroll account. When a business has multiple bank accounts, the total amount of money in each bank account is listed separately in the "Cash in Banks" category.

Accounts Receivable (Line 6) includes all money owed by guests, employees, and officers of the company, and by others such as concessionaires (for example, a business renting space from the restaurant).

Accounts receivable from Credit Card (Line 7) companies reflect the amount of an operation's credit card sales not yet paid by the operation's **merchant services provider**.

Customer House Accounts (Line 8) reflects money owed by guests to whom an operation has extended credit. The amount entered for Customer House Accounts must first be reduced by an appropriate **allowance for doubtful accounts**. If a business identifies the amount of money that is owed by others, it must also identify the reasonable amount of that debt that it is unlikely to recover. The actual amount of this allowance is based on historical experience and a realistic analysis of the business's AR balances.

Key Term
___

Merchant services provider:
A company that enables foodservice operators (merchants) to accept credit and debit card payments and other alternative payment methods. Also sometimes referred to as a "payment services provider."

Key Term
___

Allowance for doubtful accounts: A reserve held for probable losses when all AR are not collected.

What Would You Do? 4.1

"I think this could be a problem; it's going to be really close!" said Sean Kilpatrick, the manager of Santo's Charcuterie and Bakery. Sean was looking at his laptop and talking to Rita, the operation's assistant manager.

"What's going on?" asked Rita.

"Well, you know how we have payroll due every two weeks on Wednesday?" said Sean.

"Sure, I know that, so what's the problem?" replied Rita.

"Well, this month there are three payrolls due because the month has five Wednesdays. And I just paid our winter property tax bill last week," said Sean.

"OK, I still don't see the problem," said Rita.

"The problem is cash," said Sean. "business is great, but between three payrolls this month and our property tax bill being paid, I'm not sure we're going to have enough cash on hand to cover that last payroll!"

"That would be a problem for sure!" said Rita.

Assume you were the owner of Santo's Charcuterie and Bakery. How important do you think it would be that the amount of cash on hand is sufficient to cover normal fluctuations in the operation's cash needs? What would be the likely result if there was not enough cash on hand?

Inventories (Line 9) consist of the total inventory value of food, beverages, supplies, and other goods including logoed items sold by the operation (for example, coffee mugs, hats, and T-shirts). Food Inventories (Line 10) are listed first, followed by Beverage Inventories (Line 11) and Supplies and Other (Line 12).

The inventory of food and beverages include all food in storerooms, freezers, pantries, kitchens, and dry storage area, along with beverages in stock at beverage outlets (bars) and in wine cellars and dry storage areas.

Inventory values ideally should be reported at the lower of cost or current market value. However, because of the rapid turnover of inventory, many foodservice operations value inventories at their most recent specific item costs, writing the purchase cost on the box, can, or bottle in which the item is packaged when it is received.

In most hospitality industry situations, the value of an inventory item equals the amount the business paid for it. For example, when computing the value of its inventory, a restaurant would assign a case of Johnny Walker Scotch the same value as it paid for the case.

In those unique situations, the current value of an item is significantly different than the price a business paid for it (for example, extremely rare wines that may have been held in a restaurant's wine cellar for many years and have appreciated greatly in value). Then, because of the conservatism principle of accounting, the lower of the original cost or the current value should be the item's inventory value. Regardless of the inventory valuation methods used, the same procedures should be followed each accounting period.

Prepaid Expenses (Line 13) include all unexpired insurance premiums and prepaid interest, taxes, licenses, and rent. The GAAP principle of Matching (see Chapter 1) dictates that disbursements for these items be shown as assets until the operation receives their benefit; then the expense is recognized.

Generally, prepaid items are amortized (spread over) on a straight-line or equal basis. To better illustrate how foodservice operators calculate the value of prepaid expenses for balance sheet purposes, assume that a tavern owner purchases a one-year liquor liability policy that costs $12,000. The policy is paid for, in its entirety, on January 1 and a one-time payment of $12,000 was made from the tavern owner's cash account.

If the tavern owner prepared a balance sheet on February 1, the prepaid expense for liquor liability insurance would be valued at $11,000. The reason: The $12,000 policy covers a 12-month period, and the amount "used up" is $1,000 per month; it is calculated as follows:

$12,000 paid/12 months = $1,000 per month

As a result, because the value of the January payment has already been realized (one month's payment was made) on February 1, the remaining "prepaid" amount of the insurance would be $11,000, calculated as:

$12,000 per year paid − $1,000 used in January = $11,000 prepaid insurance remaining

Similarly, if the tavern's owner produced a balance sheet on the last day of July, the remaining "prepaid" amount of the insurance would be $5,000 calculated as:

$12,000 per year paid − ($1,000 × 7 months) = $5,000 prepaid insurance remaining

If, in this example, the tavern owner did not indicate the liquor liability policy had been prepaid, it would understate the value of this operator's current assets.

Total Current Assets (Line 14) is the sum of the Cash, Accounts Receivable, Inventories, and Prepaid Expenses categories.

Property and Equipment (Line 15) is a category often referred to as "Fixed Assets" (see Chapter 3). These assets are resources of the operation that are tangible, are material in amount, will be used in the operation to generate sales, and will benefit the operation for more than one year in the future.

In theory, a pen used by a foodservice operator to write notes could last more than one year. However, it would not be recognized as a fixed asset because its cost was not significant. Instead, fixed assets include items such as land, buildings, leasehold improvements, and furniture, fixtures, and equipment (FF&E).

In the Figure 4.1 example, Peggy's Restaurant does not own the land or building on which it operates so there is no entry for land or building. **Leasehold Improvements** (Line 16) include the value of modifications made to an operation's rented facilities. Leasehold improvements are actually amortized (not depreciated) because, once installed, they're technically owned by the property owner.

For example, assume a foodservice operator installs new carpet in a dining room when signing a 10-year lease, and it is expected the carpet must be replaced in five years. The cost of the

Key Term

Leasehold improvement: Anything that benefits one specific tenant in a rental arrangement. Examples include painting, adding new walls, putting up display shelves, changing flooring and lighting, and adding walls and partitions.

carpet will be amortized over that five-year period to recognize its value over the useful life of this flooring improvement.

Furniture, Fixtures, and Equipment (Line 17) includes the value of booths, tables, and chairs. These items are recorded at cost including the purchase price plus taxes, freight, insurance in transit, and installation cost.

Equipment covers a wide range of items used by a foodservice operation. Ideally, separate records are maintained for each equipment item; however, on the balance sheet these are combined. Equipment includes items such as kitchen and cleaning equipment, cash registers, and delivery vehicles.

The cost of equipment includes their purchase price, taxes on their purchase, freight charges, insurance charges during shipping, and installation charges. As with furniture and fixtures, equipment is depreciated over its useful life and the **net book value** is shown on the balance sheet. In the USAR format for the balance sheet, the value of linens, china, glassware, silver, and uniforms are also included in this line.

Key Term

Net book value: The cost of a fixed asset minus its accumulated depreciation.

In the "Less: Accumulated Depreciation and Amortization" entry (Line 18), accountants enter the cumulative depreciation of a business's fixed assets when the balance sheet is prepared. Accumulated depreciation and amortization is a contra asset account (see Chapter 2), and its natural balance is a credit that reduces the overall value of the business's property and equipment.

Total Property and Equipment—Net (Line 19) is the sum of Leasehold Improvements and FF&E, minus accumulated depreciation and amortization.

Other Assets (Line 20) can include a variety of items not previously listed on the balance sheet. Depending on the specific operation, these may include:

Rental deposits: Some foodservice leases require a cash deposit be applied to last month's rent (referred to as the rental deposit). Since the cash in this deposit is not readily available, it should be shown as part of other assets.

Organization costs: The initial cost to organize the business should be recorded as an asset and then amortized against revenue over a three- to five-year period. Organization costs can include legal fees, promotional costs, stock certificate cost, and accounting and state incorporating fees.

Preopening costs: The costs associated with opening a new business, such as payroll prior to opening and training and promotional cost, should be recorded in this account. Preopening costs are generally amortized against revenue over a one- to three-year period.

Franchise fees: This category of other assets include franchise fees if the business is a franchised operation (see Chapter 3). Franchise costs normally include an initial franchise fee and continuing royalties based on revenue. The initial fee is an "Other Asset." In addition to the franchise fee, the franchisee could also use this account for legal fees and other costs incurred to obtain the franchise. These costs will typically be amortized over the life of the franchise agreement.

Goodwill: **Goodwill** is the sum of all special advantages not otherwise identified that relate to the business. Goodwill is not recognized on the business's balance sheet unless it has been purchased. Purchased goodwill results when the cost of an acquired business exceeds the amount assigned to its individual assets. Goodwill should be reported on the balance sheet at cost less the amount written off based on impairment (loss of value of the goodwill due to factors such as greater competition or change in location).

> **Key Term**
>
> **Goodwill (Other Asset):** The sum of all special advantages relating to the business including its reputation, well-trained staff, highly motivated management, and favorable location.

In the example of Peggy's Restaurant, Other Assets include the operation's Cost of Liquor License (Line 21) and Lease and Utility Deposits (Line 22).

Total Other Assets (Line 23) is the sum of all categories listed under the Other Assets heading.

TOTAL ASSETS (Line 24) is calculated as the sum of :

✓ Total Current Assets
✓ Total Property and Equipment—Net
✓ Total Other Assets

The "Total Assets" entry on the USAR-formatted Peggy's Restaurant Balance Sheet shows the combined value of all of the business's current assets, property and equipment, and other assets. While not all balance sheets are as detailed as that recommended by USAR, it is easy to see that the methods of depreciation and amortization used by a business have a significant effect on how the value of assets is reported. This is a reason why, regardless of the format used, the methods of depreciation and amortization utilized by a business should be documented and shown in a **Notes to the Financial Statements** section.

> **Key Term**
>
> **Notes to the Financial Statements:** A supplemental narrative statement managerial accountants attach to a financial report to provide useful and critical information that would otherwise not be available.

Find Out More

Notes to the Financial Statement are intended to help readers understand and analyze the financial information included in the financial statement they are reading. These notes are designed to adequately disclose any information required to make the statements more useful and not misleading.

Examples of important information typically included in Notes to the Financial Statements include the specific accounting policies and practices used to prepare a business's income statement, balance sheet, and statement of cash flows. In addition, managerial accountants may elect to include notes that address any or all of the following:

✓ Important accounting policies
✓ Relevant changes to previously used accounting policies or methods
✓ Assets subject to a lien
✓ Lease commitments
✓ Income taxes and deferred taxes
✓ Current or pending litigation of a material nature
✓ Relevant events that have occurred since the date of the financial document's preparation

To see an example of these types of notes, enter "examples of Notes to Financial Statements" in your favorite browser and view the results.

Those who read the Assets section of a balance sheet must recognize that methods used to establish the value of a business's assets vary by asset type. Figure 4.2 summarizes methods to determine an asset's worth.

Despite differences in how asset worth is determined, the values are summed together on the balance sheet to estimate the total worth of a company's assets.

Asset Type	Worth Established By
Cash	Current value
Marketable securities	Fair market value or amortized cost
Accounts receivable	Estimated future value
Inventories	The lesser of current value or price paid
Investments	Fair market value or amortized cost
Property and equipment	Price paid adjusted for allowed depreciation
Intangible assets	Amortized value

Figure 4.2 Methods of Determining Asset Worth

However, balance sheet readers must know about calculation methods when evaluating a company's financial position and ensure they communicate the true (actual) value of the business's non-cash assets.

Liabilities and Owners' Equity

Liabilities and Shareholders' Equity (Line 25) begins to fill in the right side of the basic accounting equation:

Assets = *Liabilities and Owners' Equity*

Current Liabilities (Line 26) are listed first. A liability is current if it must be satisfied with current assets within one year.

Accounts Payable (Line 27) is often shortened to AP and reports the cost of goods and services purchased on credit (on account) by the business. Technically, AP should be recorded when legal title of property is transferred to the buyer. Practically speaking, this occurs when goods or services are delivered.

Key Term

Current liability: The total of all obligations due to be paid within 12 months of the balance sheet date.

If a specific AP is of significant size, as for example when a large piece of equipment has been purchased and installed but not yet paid for, the AP amount should be shown separately on its own line of the balance sheet.

Payroll Taxes Payable (Line 28) should include any amounts due for withholding and FICA (Social Security) taxes, and federal and state unemployment taxes. Sales Taxes Payable (Line 29) reports the amount of sales tax and any alcoholic beverage taxes currently due.

Gift Cards Payable (Line 30) is used to report the sale of gift certificates. When sold, the amount received should be credited to the Gift Cards Payable account. As gift cards are redeemed, this account is debited, and the appropriate sales account is credited. Note: It is common that a percentage of gift cards are not redeemed. This should be considered at the balance sheet date, and the account balance should reflect the value of the gift cards and certificates expected to be redeemed.

Accrued Expenses (Line 31) are expenses incurred for the period but not payable until after the balance sheet date. Accrued expenses include salaries and wages, vacation and sick leave, rent, franchise royalties, interest, real estate taxes, personal property taxes, and utilities. They also include any other accrued liabilities including customer deposits or charge tips payable. For purposes of the balance sheet these expenses are combined. However, it is desirable to show any significant amounts separately.

Current Portion, Long-Term Debt (Line 32) is the amount of long-term debt due to be paid within one year of the balance sheet date. This line reduces the amount of long-term debt by the amount of the debt payable within the year.

Total Current Liabilities (Line 33) is the sum of all current liabilities listed on the balance sheet.

Long-Term Debt, Net of Current Portion (Line 34) is the second major category of liabilities. **Long-term debt** is also referred to as a non-current liability.

Long-term debt can include:

Mortgage payable: The mortgage payable account records long-term debt when a creditor has a secured prior claim against fixed assets of the business. Note: The portion due within one year of the balance sheet date should be classified as a current liability.

Bonds payable: Bonds can be issued by a business to raise capital needed for expansion, facility renovation, and new equipment. Generally, bonds have maturity dates of five years or more. When bonds are sold, certificates are issued, and the debt should be recorded in the bonds payable account.

Deferred income taxes: When a business accounts for certain items differently for tax purposes than for reporting purposes, differences in income result. For example, a restaurant may use the straight-line method for depreciating fixed assets for financial reporting purposes, but an accelerated method of depreciation for depreciating fixed assets for tax purposes. (Note: Details regarding the various methods of depreciating fixed assets are addressed in detail in Chapter 12.).

Total Liabilities (Line 35) is the sum of total current liabilities and all non-current (long-term) liabilities.

Owner's Equity (see Chapter 2) may actually be titled differently on a balance sheet based on the type of company ownership. In Figure 4.1, the owners' equity heading used for Peggy's Restaurant is SHAREHOLDERS' EQUITY (Line 36).

Capital Stock (Line 37) is the amount of all types (common and preferred) of ownership shares that the company is authorized to issue, according to its corporate charter. Capital stock can only be issued by the company and is the maximum number of shares that can ever be outstanding.

Unlike Peggy's Restaurant, unincorporated foodservice operations are typically organized as either sole proprietorships (one owner) or as partnerships (two or more owners). If the business is a sole proprietorship, the equity section of the balance sheet consists of a single line called "(Owners Name), Capital." This account would include the initial and later capital contributions by the owner, less owner withdrawals, plus all operating results (profits or losses).

If the business is a partnership, the net worth of each individual partner is shown on the balance sheet as follows:

Net Worth

Partner 1	$xxx,xxx
Partner 2	$xxx,xxx
Partner 3	$xxx,xxx
Total Net Worth	$xxx,xxx

For sole proprietorships and partnerships, the calculation of owners' equity is rather straight-forward. For corporations, the calculations are more complex. For corporations, **common stock** is the balance sheet entry that represents the number of shares of stock issued (and owned) multiplied by the value of each share.

Thus, for example, if a company has issued 1,000,000 shares of common stock and each share has a par value (see Chapter 2) of $25.00, the total value of the company's outstanding capital stock recorded on the balance sheet would be $25,000,000 (1,000,000 shares × $25.00 per share = $25,000,000).

Key Term

Common stock: A security that represents ownership in a corporation. Holders of common stock elect the company's board of directors and can vote on corporate policies.

There are a number of ways to issue various classes of stock in addition to common stock, as well as to assessing their value for purposes of the balance sheet. Professional attorneys and managerial accountants specializing in this area can help ensure an operation's balance sheet accurately reflects the value of stocks in accordance with laws governing stock-issuing corporations.

As described in Chapter 2, Retained Earnings (Line 38) is the cumulative net (retained) earnings or profits of a company after accounting for any dividends that were paid out. As an owners' equity account, retained earnings are directly affected by the results shown on the income statement (P&L) of a business.

Retained earnings increase when new profits are generated and decrease when a company loses money or pays out dividends. In most cases, additions to retained earnings come primarily through the profits earned by a business's ongoing operations so a thorough understanding of the income statement, and the net income it shows, is important prior to an operator fully understanding balance sheet development.

Total Equity (Line 39) is the sum of the Capital Stock and Retained Earnings lines. TOTAL LIABILITIES AND SHAREHOLDERS' EQUITY (Line 40) is the sum of the amount shown on the Total Liabilities and Total Equity lines. As illustrated in Figure 4.1, the amount in the TOTAL ASSETS line will equal the amount in TOTAL LIABILITIES AND SHAREHOLDERS' EQUITY, satisfying the basic accounting equation.

If the two amounts are *not* identical, then an accounting error has been made in the preparation of the balance sheet.

Technology at Work

Foodservice operators preparing their own balance sheets often find it is helpful to use a standardized spreadsheet format. When they do, their spreadsheets can be easily saved, modified with next period accounting data, and filed electronically as well as printed in a hardcopy.

While there are a number of spreadsheet providers, Microsoft's Excel is one of the most popular. Balance sheet templates produced in Excel are also widely available at no charge.

To examine some of the no-cost Excel balance sheet spreadsheets available for downloading, enter "free balance sheet templates in Excel" in your favorite search engine and review the results.

Analysis of the Balance Sheet

There are several ways balance sheet information can be analyzed. For many foodservice operators, two important methods are:

Ratio Analysis
Vertical Analysis

Ratio Analysis

While the data included in each line on a financial statement provides valuable information, sometimes it is helpful to compare information on one line to information on another line. This process is commonly referred to as **ratio analysis**.

A **ratio** is defined as the relationship between one number and another number. Mathematically, a ratio results anytime one number is divided by another number. It is often the case that knowing the relationship between two numbers is *more* valuable than knowing the value of either (or both) of the individual numbers used to create the ratio.

Key Term

Ratio analysis: The comparison of related information, most of which is found on financial statements.

Key Term

Ratio: An expression of the relationship between two numbers; computed by dividing one number by the other number.

Foodservice operators use ratio analysis to better understand the information contained in a balance sheet and to examine, in detail, the asset, liability, and owners' equity positions of a business. Some ratios most commonly used by managerial accountants to evaluate a balance sheet are:

✓ Liquidity ratios
✓ Solvency ratios
✓ Profitability ratios

Liquidity Ratios

As previously addressed, liquidity can be based on how quickly current assets can be converted to cash. **Liquidity ratios** assess how readily current assets could be converted to cash and how many liabilities those current assets could pay off. That is important because a business may have very few current assets to convert to cash for current liabilities payments, but it may also have few current liabilities.

Alternatively, another company may have many current assets that can quickly be converted to cash. However, it may also have so many current liabilities that they exceed the cash available to pay them.

While there are several liquidity ratios, the best known is the **current ratio**.

The calculation to produce the current ratio is:

Key Term

Liquidity ratios: A category of ratios that indicate a business's ability to pay current obligations.

Key Term

Current ratio: The liquidity ratio that compares current assets to current liabilities and indicates an operation's ability to pay its current obligations in a reasonable time.

$$\frac{\text{Current assets}}{\text{Current liabilities}} = \text{Current ratio}$$

For example, if an operation has current assets of $200,000 and current liabilities of $150,000, the operation's current ratio would be calculated as:

$$\frac{\$200,000}{\$125,000} = 1.6$$

In this example, the operation has assets 1.6 times its current liabilities. In other words, it has $1.60 in readily available current assets to pay each $1.00 of its current obligations (liabilities). The higher the current ratio, the more current assets a business has available to cover its current liabilities. Current ratios can be interpreted in the following manner:

When current ratios are:

Less than 1.00: The business may have a difficult time paying its short-term debt obligations because of a shortage of current assets.

Equal to 1.00: The business has an equal amount of current assets and current liabilities.

Greater than 1.00: The business has more current assets than current liabilities and should be in a good position to pay its bills as they become due.

It might seem desirable for every foodservice operation to have a high current ratio so it can easily pay all current liabilities. However, this is not always the case. While potential creditors would certainly like to see a business in a good position to pay all its short-term debts, the current ratio may be high because of significant amounts of AR.

That situation may mean that the business's credit policies (or its AR collection efforts) should be re-evaluated and improved. The current ratio is so important to the long-term viability of a business that lenders might require that any business seeking a loan must maintain, during the life of a loan if granted, a minimum current ratio level established by the lender.

Working capital relates to the current ratio. Although not a true ratio (it does not require that one number be divided by another number), it is a measure many lenders assess and require for those seeking loans.

Key Term

Working capital: The capital of a business used in its day-to-day operations.

The calculation used to determine working capital is:

Current Assets – Current Liabilities = Working Capital

For example, if an operation has current assets of $200,000 and current liabilities of $125,000, the operation's working capital is:

$$\$200,000 - \$125,000 = \$75,000$$

In this example, the operation has $75,000 in working capital. Although working capital can be calculated for all foodservice operations, new businesses are usually required by lenders to obtain a target working capital to receive and maintain start-up loans. The reason: Lenders want assurance that new businesses have enough resources to pay current obligations during the uncertainty and risk of the initial years of operations.

Key Term

Solvency: The ability of a business to meet its long-term financial obligations as they become due.

Solvency Ratios

Solvency refers to the ability of a business to meet its financial obligations as they become due.

Solvency ratios include the **solvency ratio**, the debt to equity ratio, and several others.

A business is solvent when its assets exceed its liabilities. Assets in the form of cash are received from stockholders, from successful operations, and by borrowing from lenders. The risk of insolvency increases when an operation relies more heavily on debt financing than financing provided by stockholders and internally generated financing (profit generation).

Key Term

Solvency ratio: A ratio that measures total assets against total liabilities.

The calculation used to produce the solvency ratio is:

$$\frac{\text{Total assets}}{\text{Total liabilities}} = \text{Solvency ratio}$$

For example, if an operation has total assets of $1,360,000 and current liabilities of $850,000, the operation's solvency ratio is calculated as:

$$\frac{\$1,360,000}{\$850,000} = 1.6$$

In this example, the business has $1.60 of total assets to every $1.00 of total liabilities. To better understand this ratio, consider it is a comparison between what a company "owns" (its assets) and what it "owes" to those who do not own the company (its liabilities).

A solvency ratio of "1.00" would mean that assets = liabilities. Values over 1.00 indicate that a company owns more than it owes, and values less than 1.00 indicate a company owes more than it owns. Creditors and lenders prefer to do business with companies that have a high solvency ratio (between 1.5 and 2.0) because these companies are likely to have the ability to repay their debts.

Key Term

The **debt to equity ratio** contrasts the amount of money that creditors have loaned the business to the number of dollars owners have invested in the business.

Debt to equity ratio: A ratio that compares a business's equity to its total liabilities.

The debt to equity ratio is used by managerial accountants to evaluate the relationship between investments made by an operation's lenders and investments made by the operation's owners.

The calculation used to produce the debt to equity ratio is:

$$\frac{\text{Total liabilities}}{\text{Total owners' equity}} = \text{Debt to equity ratio}$$

For example, if a business has total liabilities of $750,000, and total owners' equity of $1,000,000, the business's debt to equity is calculated as:

$$\frac{\$750,000}{\$1,000,000} = 0.75 \text{ times}$$

In this example, the business has $0.75 of total liabilities to every $1.00 of total owners' equity.

From a lender's perspective, the higher the lender's own investment relative to the actual investment of the business's owners, the riskier the investment and the less they are likely to be interested in loaning money to the business.

Many restaurants are risky investments, and lenders are typically cautious about loaning money to them. The result is that lenders usually look favorably on projects that yield relatively low (less than 1.00) ratios of this type.

All other things equal, a company that can generate and retain earnings (as part of owners' equity) will be able to reduce this ratio and reduce risk to potential lenders.

Profitability Ratios

In the final analysis, it is the job of management to generate profits for a company's owners, and **profitability ratios** measure how well management has accomplished this task.

Profitability ratios help operators better understand the "bottom line" (actual profitability) of companies they manage or analyze. A company's profitability cannot be properly measured only by evaluating the amount of money it returns to the company's owners. Profits must also be evaluated based on the size of investment in the business made by the company's owners. The higher the investment of the company's owners, the greater will be their profit expectations.

There are several profitability ratios of interest to balance sheet readers. Two important profitability ratios are:

✓ Return on assets (ROA)
✓ Return on owner's equity (ROE)

Return on Assets

When investors provide funds or secure assets for a business, they hope the money provided will, at some point, be returned to them, and they also seek profits. The original investment and profits paid back to owners are called **financial returns**.

Key Term

Profitability ratios: A group of ratios that assess a business's ability to generate earnings relative to its revenue, operating costs, balance sheet assets, or shareholders' equity.

Key Term

Financial returns: The money made or lost on an investment over a specific time.

When utilizing profitability ratios, managerial accountants want to examine the size of an investor's financial return. **Return on assets (ROA)** is one ratio that helps them do this.

Return on assets (ROA) shows a business's ability to use its total assets to generate net income. To calculate this ratio, operators use information from both the balance sheet (Total assets) *and* the income statement (Net income) examined previously.

Key Term

Return on assets (ROA): A financial ratio indicating the profitability of a company in relation to total assets.

The calculation used to produce the ROA ratio is:

$$\frac{\text{Net income}}{\text{Total assets}} = \text{Return on assets (ROA)}$$

For example, if a business operated for one year and had net income of $150,000 and total assets of $1,000,000, the business's ROA for that year would be:

$$\frac{\$150,000}{\$1,000,000} = 0.15 \text{ or } 15\%$$

In this example, the business made $0.15 of net income for every $1.00 of total assets.

This ratio is important and easy to understand if you analyze it by comparing two businesses. Assume that two foodservice operations generate a net income of $100,000 per year. If the owners of Operation A commit assets of $1,000,000 to operate their property, and the owners of Operation B commit assets valued at $10,000,000, the rate of financial return (relative to the value of their assets) is very different.

Operation A

$$\frac{\$100,000}{\$1,000,000} = 0.10 \text{ or } 10.0\% \text{ return on assets}$$

Operation B

$$\frac{\$100,000}{\$10,000,000} = 0.01 \text{ or } 1.0\% \text{ return on assets}$$

In this example, the ROAs are very different, despite each operation generating the same net income.

A ROA of over 5% is generally considered good (20% is excellent) and, not surprisingly, investors seek high rates of ROA. Like nearly all investments, however, when returns in a restaurant are perceived to be higher, the risks associated with these investments also tend to be higher.

Lower rates of return are often associated with lower risk. ROA is dependent upon net income achieved through operations (the numerator in the equation). Rates of ROAs can be low because the business using the assets has a low net income because high depreciation or interest expenses were removed.

A low ROA also could result from excess assets (the denominator in the equation). For these reasons, managerial accountants carefully review the ROA achieved by a business and compare it to industry averages, the business owner's own investment goals, and other valid **benchmarks**.

Return on Owners' Equity

The **return on equity (ROE)** ratio was developed to evaluate the rate of return on the personal funds invested by the owners (and/or shareholders) of a business.

Like ROA, to generate the ROE ratio, operators use information from both the balance sheet (total owners' equity) *and* the income statement (net income).

The calculation used to produce the ROE ratio is:

Key Term

Benchmark: Something serving as a standard by which others may be measured or judged.

Key Term

Return on equity (ROE): The measure of a company's net income divided by its shareholders' equity. ROE is a gauge of a corporation's profitability and how efficiently it generates those profits. The higher the ROE, the better a company is at converting its equity financing into profits.

$$\frac{\text{Net income}}{\text{Total owners' equity}} = \text{Return on equity (ROE)}$$

For example, if a business operated for one year and had net income of $70,000 and total owners' equity of $1,000,000, the business's ROE for that year would be calculated as:

$$\frac{\$70,000}{\$1,000,000} = 0.07 \text{ or } 7\%$$

In this example, the business made $0.07 of net income for every $1.00 of owners' investment.

ROEs of 15–20% are generally considered to be good. However, the ways investors' rates of return are computed can vary. Some managerial accountants believe returns based on pre-tax income are most informative, while others compute rates of return after all taxes are paid. Both methods have value.

An "after taxes" return on owners' equity is usually computed because only the after-tax income is available to corporations for a dividend return to shareholders.

Even for individual business owners, the greatest interest is in determining the total amount personally invested and the total amount (after taxes) that investment will produce (return to them) each year.

If owners knew there was no risk of losing their initial cash contribution investments, they would always desire to maximize the size of their return on owners' equity. However, in the business world, increased returns are typically associated with increased investment risk. Most experienced investors desire to achieve their own preferred balance between the risk they are willing to assume, and the investment returns they desire to achieve.

Find Out More

In addition to the financial ratios described in this chapter, managerial accountants have created many other ratios to help them analyze information on their business's balance sheets, income statements (previous chapter), and statements of cashflows (addressed in the next chapter).

In several cases, the ratios take numbers from more than one of these financial statements to provide insightful information.

For example, a ratio such as earnings per share (EPS) is calculated as:

$$\frac{\text{Net income}}{\text{Average number of common shares outstanding}} = \text{EPS}$$

This calculation requires an operator to use a number from the income statement (net income) and a number from the balance sheet (average number of common shares outstanding). This ratio may have great meaning to corporations that issue common stock to stockholders. Corporations whose stock is not publicly traded will have little or no interest in this ratio. Similarly, this is not á ratio that would be of interest to a sole proprietor. Foodservice operators should utilize the financial statement ratios most meaningful to their own businesses.

To see additional examples of financial ratios commonly used by owners and operators in the foodservice industry, enter "examples of financial statement ratios used by restaurants" in your favorite browser and view the results.

Vertical Analysis

In most cases, it is not possible to directly compare the information on the balance sheet of one company to that of another company. One reason relates to differences in company sizes and how different businesses are financed. However, one method managerial accountants use to make some balance sheet (and other financial statements) comparisons is by **vertical analysis**.

When using vertical analysis on a balance sheet, the business's Total Assets are given a value of 100% and Total Liabilities and Owners' Equity also take a value of 100%. This can easily be seen in Figure 4.3 in the shaded "%"column head inserted in the balance sheet (Line 24) where Peggy's Restaurant Total Assets equal $533,400 (100%) and Total Liabilities and Owners' Equity equals $533,400 (100% in Line 40).

Key Term

Vertical analysis: A method of financial statement analysis in which each line item is listed as a percentage of a base figure within the statement. Also known as "common-size" analysis.

When utilizing vertical analysis, individual asset categories are expressed as a percentage (fraction) of Total Assets. Individual liability and owners' equity classifications are expressed as a percentage of Total Liabilities and Owners' Equity (as shown in Figure 4.3). Therefore, each percentage computed is a percent of a "common" number, which is why this type of analysis is sometimes referred to as "common-size" analysis.

Peggy's Restaurant

Balance Sheet (Vertical Analysis)

As of December 31, 20xx

Line		$	$	$	%
1	**Assets**				
2	**CURRENT ASSETS**	$	$	$	%
3	Cash:				
4	House Banks	$6,500			
5	Cash in Banks	35,800			
6	Accounts Receivable:		42,300		7.9%
7	Credit Cards	13,800			
8	Customer House Accounts	8,500			
9	Inventories:		22,300		4.2%
10	Food	18,000			
11	Beverage	15,600			
12	Supplies & Other	7,500	41,100		7.7%
13	Prepaid Expenses		18,500		3.5%
14	**Total Current Assets**			124,200	23.3%

Figure 4.3 USAR Balance Sheet Vertical Analysis

Peggy's Restaurant

Balance Sheet (Vertical Analysis)

As of December 31, 20xx

Line			
15	**PROPERTY AND EQUIPMENT**		
16	Leasehold Improvements	321,400	
17	Furniture, Fixtures, and Equipment	133,000	
18	Less: Accumulated Depreciation and Amortization	−68,500	
19	**Total Property and Equipment—Net**		385,900	72.3%
20	**OTHER ASSETS**		
21	Cost of Liquor License	15,800	
22	Lease and Utility Deposits	7,500	
23	**Total Other Assets**		23,300	7.6%
24	**TOTAL ASSETS**		**$533,400**	**100%**
25	**Liabilities and Shareholders' Equity**		
26	**CURRENT LIABILITIES**		
27	Accounts Payable	38,600	
28	Payroll Taxes Payable	21,900	
29	Sales Taxes Payable	27,800	
30	Gift Cards Payable	11,500	
31	Accrued Expenses	10,000	
32	Current Portion, Long-Term Debt	27,600	
33	**Total Current Liabilities**		137,400	25.8%
34	Long-Term Debt, Net of Current Portion		168,000	31.5%
35	**Total Liabilities**		305,400	57.3%
36	**SHAREHOLDERS' EQUITY**		
37	Capital Stock	35,000	
38	Retained Earnings	193,000	
39	**Total Equity**		228,000	42.7%
40	**TOTAL LIABILITIES AND SHAREHOLDERS' EQUITY**		**$533,400**	**100%**

Figure 4.3 *(Continued)*

A balance sheet summarizes a foodservice business's financial position at a single point in time. As was true with the income statement addressed in the previous chapter, while a properly prepared balance sheet is an accurate summary of a business's assets, liabilities, and owners' equity, it makes no declaration about the business itself.

Those who read and analyze the balance sheet typically do so to determine the financial strength of a business. Information on the balance sheet helps assess a company's ability to pay for its near-term operating needs, meet future debt obligations, and make distributions to owners.

To help balance sheet readers determine the quality of an operation's performance, they can also compare a business's vertical analysis percentages to four different benchmark sources of information:

✓ Similar businesses
✓ Industry averages
✓ Financial position shown on earlier balance sheet
✓ Budgeted (planned) financial position

Technology at Work

Sometimes, it is confusing to know what to read while analyzing financial information on a balance sheet. This is especially true for those who learn or understand best visually or by hearing spoken information.

Balance sheet analysis is an area of managerial accounting for which you can find instructional videos on YouTube. The videos provide different lengths, detail, and complexity but when carefully chosen they can provide a wealth of free information.

To view some entertaining and informative videos addressing balance sheets and how to interpret them, enter "how to analyze a balance sheet" in the YouTube search bar and view the results.

The income statement and balance sheet are two of the three key financial statements required to be produced when a business utilizes GAAP. The Statement of Cash Flows is the final of these three required financial statements and its production and use are the topics of the next chapter.

What Would You Do? 4.2

Larry and Mary are two foodservice operators and long-time friends who own food trucks. While attending their local restaurant association's monthly meeting, they had a conversation over lunch.

(Continued)

"Well," said Larry, "we had a really great year. We had sales of $350,000, and we achieved a net income of $70,000!"

"We had a great year too," said Mary, "we had $450,000 in revenue and we made $100,000 in net income."

"Well, we didn't make that much," said Larry, "but because my truck was used when I bought it, and only cost me $80,000, I think I did pretty good."

Assume you were Mary and you believed Larry was honest about his numbers. Why would the amount of money Larry invested in his truck affect how his profits should be viewed? Would you need to make the same type of investment considerations as you evaluated the financial performance of your own food truck? Why or why not?

Key Terms

Cash on hand	Goodwill (Other Asset)	Working capital
Liquidity	Notes to the Financial	Solvency
Cash flows	Statements	Solvency ratio
Lien	Current liability	Debt to equity ratio
Petty cash	Long-term debt	Profitability ratios
Merchant services	Capital stock	Financial returns
provider	Common stock	Return on assets (ROA)
Allowance for doubtful	Ratio analysis	Benchmark
accounts	Ratio	Return on equity (ROE)
Leasehold improvement	Liquidity ratios	Vertical analysis
Net book value	Current ratio	

Operator's 10-Point Tactics for Success Checklist

Evaluate your need for, and the current status of, each of the following operational tactics. For those tactics you think are important, but not yet in place, develop an action plan for its implementation including who will be responsible for the tactic's completion and the target date by which it should be completed.

Tactic	Don't Agree (Not Done)	Agree (Done)	Agree (Not Done)	Who Is Responsible?	Target Completion Date
				If Not Done	
1) Operator understands the importance of the balance sheet in reporting the financial health of a business.	——	——	——		

Tactic	Don't Agree (Not Done)	Agree (Done)	Agree (Not Done)	If Not Done	
				Who Is Responsible?	Target Completion Date
2) Operator understands the key differences in the interests of the various users of the balance sheet.	——	——	——		
3) Operator recognizes the limitations associated with assessing the financial health of a business when reading a balance sheet.	——	——	——		
4) Operator knows the layout of a balance sheet prepared using the USAR.	——	——	——		
5) Operator understands the content of the Assets portion of a balance sheet.	——	——	——		
6) Operator understands the content of the Liabilities, and Owners' Equity portion of a balance sheet.	——	——	——		
7) Operator recognizes the value of liquidity ratios when analyzing a balance sheet.	——	——	——		
8) Operator recognizes the value of solvency ratios when analyzing a balance sheet.	——	——	——		
9) Operator recognizes the value of profitability ratios when analyzing a balance sheet.	——	——	——		
10) Operator understands the method used to apply vertical analysis to a balance sheet.	——	——	——		

5

The Statement of Cash Flows

What You Will Learn

1) The Importance of the Statement of Cash Flows
2) How Sources and Uses of Funds Are Reported on the Statement of Cash Flows
3) How to Analyze a Statement of Cash Flows

Operator's Brief

In business, it is often said that "cash is king." Those making this statement refer to the absolutely critical role that *cash* and *cash management* play in a successful business operation.

In this chapter, you will learn that businesses producing cash in excess of their immediate needs have a "positive cash flow: more cash is flowing *into* the business than is being removed from it." In contrast, businesses that do not generate enough cash to support their operations have a "negative cash flow."

It is possible for a business with a negative cash flow to operate for some time. However, unless cash is provided from another source, a business with a consistent negative cash flow will run out of the money (cash) needed to pay bills as they come due.

The Statement of Cash Flows (SCF) is the financial report that tells its readers about increases (inflows) and decreases (outflows) in the business's cash management.

Typically, businesses accumulate cash by making sales that exceed their spending of cash for necessary expenses. They may invest their excess cash or use it to pay down debts the business may have. Shortfalls in cash are made up by borrowing money or by acquiring funds provided by investors.

The SCF examines the cash flows resulting from a business's operating, investing, and financing activities. In this chapter, you will learn about the sources and uses of funds that affect cash availability. You will also learn how managers create an SCF. Finally, you will discover how foodservice operators can read, analyze, and utilize the important information contained in an SCF.

CHAPTER OUTLINE

The Importance of Cash Flow
Sources and Uses of Funds
Creating the Statement of Cash Flows
    Cash Flow from Operating Activities
    Cash Flow from Investing Activities
    Cash Flow from Financing Activities
    Net Changes in Cash
    Supplementary Schedules
Statement of Cash Flows Analysis

The Importance of Cash Flow

The proper management of **cash** as it flows through a foodservice operation is critical. Cash is any medium of exchange readily accepted by a bank. Cash can be easily transferred and, therefore, is the one foodservice business asset most likely to be improperly used (stolen) by staff members. Insufficient cash can lead to **bankruptcy**, and excess cash not properly invested is a resource being wasted.

Because cash is so important, effective foodservice operators utilize a variety of tactics to ensure their **cash management** systems work properly.

For many years, businesses that issued income statements (see Chapter 3) and balance sheets (see Chapter 4) to those outside the company were encouraged to also supply a document called the "Statement of Changes in Financial Position," also known as the "Funds Statement." The intent of this statement was to indicate how cash inflows and outflows affected a business during a specific accounting period reported on an income statement.

Key Term

Cash: Any medium of exchange a bank will accept at face value.

Key Term

Bankruptcy: A legal proceeding initiated when a person or business is unable to repay outstanding debts or obligations.

Key Term

Cash management: The process used to effectively control a business's cash balances (currency on hand and demand deposits in banks), cash flow (cash receipts and disbursements), and short-term investments in securities.

In 1961, the American Institute of Certified Public Accountants (AICPA) officially recommended that a funds statement be included with the income statement and balance sheet in annual reports to shareholders. In 1971, the Accounting Principles Board (APB) officially added the funds statement to yield the three primary financial documents required in an annual report to shareholders.

In 1987, the Financial Accounting Standards Board (FASB) called for a statement of cash flows to replace the more general funds statement. The **Statement of Cash Flows (SCF)** shows all sources and uses of funds from operating, investing, and financing activities of a business for a specified accounting period.

As its name implies, cash flows from operating activities represent the amount of cash generated by a business's basic operations. For foodservice operators, these include transactions involving acquiring and selling food and beverage products and providing guest services. Cash flows from operating activities include money collected from customers for their purchases, cash paid to employees and other suppliers, interest and taxes paid, and other operating payments.

Key Term

Statement of Cash Flows (SCF): A report providing information about all cash inflows a company receives from its ongoing operations and external investment sources. It also includes all cash outflows paying for business activities and investments during a given accounting period.

Cash flows from investing activities reflect changes in cash resulting from transactions related to asset accounts not directly affecting day-to-day operations. These transactions typically include the acquisition and disposal of property and facilities and the purchase and sale of investments in other businesses.

Cash flow from financing activities reflect cash changes related to various liability and equity accounts shown on the balance sheet and which do not directly affect the business's operation. Examples include obtaining and repaying debt, issuing and repurchasing stock, and dividend payments to company owners.

It is important to recognize that the FASB, in efforts to help investors and creditors better predict future cash flows, has specified a recommended universal format for producing an SCF. The Uniform System of Accounts for Restaurants (USAR) recommends the use of SCF in formats appropriate for their users and based directly on FASB recommendations.

Why do the FASB and the International Financial Reporting Standards (IFRS) now require the use of an SCF? Consider Rodney Decker, who owns and operates three nationally franchised coffee shops in his town. Each month, Rodney and his accountant prepare an individual income statement for each coffee shop and a

consolidated income statement combining revenue, expense, and profit information from each of his individual coffee shops into one (overall summary) income statement.

In addition, each month they prepare an updated balance sheet for Rodney's business. As addressed in the two previous chapters, Rodney can learn much about his business by reviewing the information on his consolidated income statement and balance sheet. What he could *not* learn from studying them, however, includes:

✓ How could my current assets have decreased if I made a profit in the accounting period?
✓ Do I have cash available for purchasing additional investments like stocks and bonds?
✓ Do I have cash available if I want to expand my business, or will I need to borrow money to do so?
✓ If I take money out of my business in the form of profits paid to me, will the business have enough remaining cash to meet its operational needs?
✓ Can I take cash out of my business to go on a vacation even if the business lost money during the accounting period?

In this example, there are a variety of questions Rodney might like to address that cannot be answered with only the information found on his income statement and/or balance sheet.

The cash inflows and outflows of a business are significant to a business's owners, investors, lenders, creditors, and managers. The presentation of accurate information about cash flows should enable investors in a business to:

1) Predict the amount of cash to be distributed with profit payouts or dividend distributions.
2) Evaluate the possible risk associated with continued investment in the business.

The previous chapter introduced the concept of "liquidity," the ease with which a current asset can be converted to cash. Therefore, liquidity can also be considered as "nearness to cash." In a similar manner, solvency (see Chapter 4) refers to the business's abilities to pay its debts as they become due. Since most debts incurred by a business are repaid in cash, solvency is an important measure of a firm's likelihood to remain a **going concern**.

If a business is *not* considered by lenders and investors to be a going concern, the business will likely find its ability to borrow money is severely restricted or diminished. If it can borrow money, the funds will come with the increased costs that reflect the higher risk associated with lending money to a business that does not consistently demonstrate a strong positive cash flow. The reason: When cash flows are not positive, investors demand higher ROI levels to compensate for the greater risk they are taking. Lenders will also desire higher interest rates to compensate themselves for that same higher risk level.

Technology at Work

The timely and accurate financial reporting on businesses with multiple business units in multiple locations can be complex. Some businesses must blend the income statements from dozens or even hundreds of individual operations to arrive at a properly prepared consolidated income statement.

Fortunately, many companies offer multi-entity accounting software that helps consolidate financial data from multiple sources.

To examine some software options available to foodservice operators in this key area, enter "financial consolidation software" in your favorite search engine and review the results.

When properly created, the SCF Rodney prepares will report the movement of cash into and out of his business during a specific time. As addressed in Chapter 4, cash is reported on a balance sheet as a current asset. Cash, in this case, refers to currency, checks on hand, and deposits in banks. Before examining an SCF, it is important for managers to understand that cash is *not* the same as "revenue" or "sales."

To illustrate, assume that a restaurant hosted a large rehearsal dinner for a soon-to-be-married couple. The event included a reception and dinner for 100 guests. If the event's host paid the charges with a credit card, the restaurant would have made a sale (created revenue), but it would not have an immediate increase in its cash position. In fact, the operation will not actually have an increase in its cash position until its merchant services provider (see Chapter 4) deposits the money owed (minus agreed-upon processing fees!) in the operation's bank account. This is a process that typically takes several days. As a result, the restaurant would have increased its revenue, but not its cash, on the day of the rehearsal dinner.

In addition to understanding the nature of cash, operators must also understand **cash equivalents.** Rodney will use cash to pay his

Key Term

Cash equivalent: A short-term, temporary investment such as treasury bills, certificates of deposit, or commercial paper that can be quickly and easily converted to cash.

operating expenses, repay loans, and make investments. The SCF is designed to report his business's **sources and uses of funds** and beginning and ending cash and cash equivalents balances for each accounting period.

Key Term

Sources and uses of funds: Inflows and outflows of money affecting a business's cash position.

Find Out More

Whether you are a manager, business owner, entrepreneur, or investor, knowing how to read and understand an SCF can enable you to learn important information about the financial health of a company.

For investors, cash flows can help them better understand whether they should invest in a company. For a business owner or entrepreneur, cash flows can help them understand business performance and adjust key initiatives or strategies, as necessary.

Not everyone, however, has finance or accounting expertise. For non-financial professionals such as some managers and chefs, understanding the concepts behind an SCF and other financial documents can be challenging.

Not surprisingly, YouTube posts a significant number of videos in which the presenter explains the importance of the SCF and how to interpret them. To view some of these instructional videos, go to YouTube. When you arrive at their site, enter "how to read and interpret a statement of cash flows" in the search bar and view some of the results.

Sources and Uses of Funds

The preparation of an SCF can better be understood by first understanding the inflow and outflow of business funds.

As shown in Figure 5.1, sources of funds represent *inflows* and uses of funds represent *outflows*. These flows can be observed when balance sheets from two different accounting periods are examined. The reason: When comparing asset values from a previous period's balance sheet to a current period balance sheet, *increases* in assets represent uses of funds, and *decreases* in assets represent sources of funds.

The effects of value changes on selected asset accounts commonly used by restaurants are summarized in Figure 5.2. Note that accumulated depreciation behaves in an opposite manner to the other asset accounts. This is

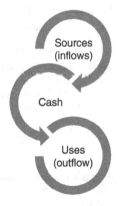

Figure 5.1 Sources and Uses of Cash

Assets	Increases	Decreases
Current Assets		
Cash on Hand and in Banks	Use	Source
Accounts Receivable (Net)	Use	Source
Food Inventory	Use	Source
Beverage Inventory	Use	Source
Supplies Inventory	Use	Source
Prepaid Expenses	Use	Source
Short-Term Investments	Use	Source
Property and Equipment		
Land	Use	Source
Leasehold Improvements	Use	Source
Furniture, Fixtures, and Equipment (FF&E)	Use	Source
Tableware and Kitchen Utensils	Use	Source
Accumulated Depreciation	Source	Use
Other Assets	Use	Source

Figure 5.2 Sources and Uses of Funds for Common Foodservice Operation Assets

because (see Chapter 2) depreciation is a contra asset account that *decreases* the balance in an asset account. The account is not classified as an asset since it does not represent a long-term value. Likewise, it is not considered a liability since it does not constitute a future obligation.

To better understand uses and sources of funds affecting a business's balance sheet, consider the direct effect of increasing or decreasing each of a foodservice operation's typical asset accounts shown in Figure 5.2 using the examples provided.

Asset Account	Source or Use	Example
Decrease in Cash on Hand and in Banks	Source	Cash on the balance sheet represents an operation's account(s) balances in the bank. When money is withdrawn from the bank, it decreases the level of the cash account, and the cash that was withdrawn becomes cash in hand (available to be spent).
Increase in Cash on Hand and in Banks	Use	When cash is deposited in the bank, it creates an increase in the cash account. The result is less cash in hand that can be spent. In effect, the bank is "using" the operation's money, so it represents a use of funds.

Asset Account	Source or Use	Example
Decrease in Net Receivables	Source	Accounts receivable represents money that guests owe an operation for products or services already provided. If guests pay what they owe, this decreases accounts receivable and provides a source of funds for the operation.
Increase in Net Receivables	Use	Accounts receivable represents money that guests owe the operation. If accounts receivable increases, the operation is owed more money than before. Since the operation has already provided products or services without being paid, this is a use of funds.
Decrease in Food, Beverage, and Supplies Inventories	Source	If inventory (i.e., food and beverage products and the supplies needed to make them) is sold to guests and they pay for it, this is a source of funds. Selling the products decreases inventory levels.
Increase in Inventories	Use	When an operator buys more inventory from its suppliers, it is a use of funds. Buying the products increases inventory values.
Decrease in Prepaid Expenses	Source	Prepaid expenses are items to be used within a year's time, but which must be completely paid for at the time of purchase. As the prepaid expense is amortized throughout the year, it decreases. Since it was prepaid, it is a source of funds.
Increase in Prepaid Expenses	Use	If prepaid expenses increase, an item was paid in advance and is a use of funds.
Decrease in Short-Term Investments	Source	Short-term investments are marketable securities such as stocks and bonds issued by other companies and that a business owns and that the business plans to keep only for a short term (less than a year). If these securities are sold, the money received from the sale is a source of funds. This decreases the amount of short-term investments owned.
Increase in Short-Term Investments	Use	If an operation buys securities, it pays money for the purchase (a use of funds). This increases the amount of short-term investments owned.
Decrease in Operating Equipment	Source	If property or equipment is sold, it is a source of funds. Selling the property or equipment decreases this account.

(Continued)

Asset Account	Source or Use	Example
Increase in Operating Equipment	Use	If property or equipment is purchased, it is a use of funds, and buying the property or equipment increases this account.
Decrease in Land, Leasehold improvements, FF&E, Tableware, and Kitchen Utensils	Source	If land, leasehold improvement items, FF&E, tableware, or kitchen utensils are sold, it is a source of funds and selling these items decrease these accounts.
Increase in Property and Equipment	Use	If land, leasehold improvement items, FF&E, tableware, or kitchen utensils are purchased, it is a use of funds. Buying these items increases these accounts.
Increase in Accumulated Depreciation	Source	Depreciation is subtracted on the income statement to lower taxable income and associated taxes. If an operation increases the amount of depreciation subtracted, its income taxes are decreased, and provides a source of funds.
Decrease in Accumulated Depreciation	Use	If the amount of depreciation subtracted is reduced, taxes are increased, which results in a use of funds.
Decrease in Other Assets	Source	An example of another asset is restricted cash (cash deposited in a separate account for a restricted purpose such as retiring long-term debt). When money is withdrawn from this account, it increases cash on hand and is a source of funds.
Increase in Other Assets	Use	When cash is deposited into this account (increase in the restricted cash account), less cash is on hand to spend. The bank is again "using" the operation's money, so it is a use of funds.

Increases in liabilities and owners' equity represent sources of funds and decreases in liabilities and owners' equity represent uses of funds as shown in Figure 5.3.

Liabilities and Owners' Equity	Increases	Decreases
Current Liabilities		
Accounts Payable	Source	Use
Notes Payable	Source	Use
Other Current Liabilities	Source	Use
Long-Term Liabilities		
Long-Term Debt	Source	Use
Owners' Equity		
Common Stock	Source	Use
Paid-in Capital	Source	Use
Retained Earnings	Source	Use

Figure 5.3 Sources and Uses of Funds for Common Foodservice Operation Liabilities and Owners' Equity Accounts

To illustrate information in Figure 5.3, consider the direct effect of increasing or decreasing each liability and owners' equity account in the following examples.

Liabilities and Owners' Equity	Source or Use	Example
Increase in Accounts Payable	Source	Accounts payable represents money owed to suppliers for products or services already received. If accounts payable increases, an operation received products or services but did not pay for them, and this is a source of funds.
Decrease in Accounts Payable	Use	Since accounts payable represents money owed suppliers for products or services already received, when an operation pays what it owes this decreases accounts payable and provides a use of funds.
Increase in Notes Payable	Source	Notes payable are short-term (less than a year) loans. If an operation increases notes payable by borrowing money, this is a source of funds.
Decrease in Notes Payable	Use	If notes payable is decreased by paying money that was borrowed, this is a use of funds.
Increase in Other Current Liabilities	Source	Accrued wages are owed to employees for work they have already done (but have not been paid due to the timing of work and the date of the balance sheet). If an operation has benefited from their work output but have not paid employees, this is a source of funds.

(Continued)

Liabilities and Owners' Equity	Source or Use	Example
Decrease in Other Current Liabilities	Use	If an operation pays workers their wages for work previously done, this decreases other current liabilities and is a use of funds.
Increase in Long-Term Debt	Source	Long-term debt includes long-term (to be repaid in more than a year). If an operation increases long-term debt by borrowing money, this is a source of funds.
Decrease in Long-Term Debt	Use	If long-term debt is decreased because an operation pays back money it owed, this is a use of funds.
Increase in Common Stock and Paid-in Capital	Source	Common stock and paid-in capital represent company stocks sold to stockholders. If stocks are sold, a company receives money from the sale (a source of funds). This increases the amount of issued stocks shown on the balance sheet.
Decrease in Common Stock and Paid-in Capital	Use	If a company buys back stocks previously issued, this is a use of funds.
Increase in Retained Earnings	Source	Retained earnings represent the accumulated profits over the life of the business that have not been distributed as dividends. If retained earnings increases, the business made a profit and generated net income for the year, and this is a source of funds.
Decrease in Retained Earnings	Use	If retained earnings decreases, the business may have experienced an operating loss, or paid out funds to its owners, and this requires the use of funds.

Sources and uses of funds follow a pattern. Specifically, asset accounts all behave the same (with the exception of the accumulated depreciation account), liability accounts all behave the same, and owners' equity accounts all behave the same as shown in Figure 5.4 .

In Figure 5.4, up (↑) arrows represent "increases" and down (↓) arrows represent "decreases." Assets in each column have opposite arrows from liabilities and owners' equity. Also, note that arrows in the left column are opposite of those in the right column. For example, if an operator sold a piece of equipment, then the asset arrow is down, and equipment's selling price is a source of funds.

Sources	Uses
↓ Assets*	↑ Assets*
↑ Liabilities	↓ Liabilities
↑ Owners' Equity	↓ Owners' Equity

* Note: Depreciation is a contra asset account and behaves oppositely of all other asset accounts, so an ↑ in depreciation is a source and a ↓ in depreciation is a use.

Figure 5.4 Sources and Uses of Funds Summary

Creating the Statement of Cash Flows

The SCF should be prepared just as often as a business prepares its income statement and balance sheets. To create an SCF for this year's results, a managerial accountant must have:

✓ Income statement for this year (including a statement of retained earnings)
✓ Balance sheet from this year
✓ Balance sheet from last year

To illustrate the SCF development process, a Condensed Income Statement and Statement of Retained Earnings for the Blue Bayou Restaurant for this year is shown in Figure 5.5. A **condensed income statement** reports an operation's revenues, expenses, and profits in a summary format.

The condensed income statement is sufficient for the purposes of preparing an SCF. In addition, a **statement of retained earnings** reports the changes in retained earnings (accumulated amount of profits over the life of the business that have not been distributed as dividends) from last year to this year.

An example of how the Blue Bayou restaurant's statement of retaining earnings would be calculated is shown in Figure 5.6.

After developing a condensed income statement and the statement of retained earnings, managerial accountants can begin the SCF creation process. Last year's balance sheet and this

Key Term

Condensed income statement: An income statement that reduces much of the normal income statement detail to just a few lines. Typically, all revenue line items are aggregated into a single line item, while the cost of sales appears as one line item. Labor and all other operating expenses appear in separate lines.

Key Term

Statement of retained earnings: A financial document that reports the changes in a business's retained earnings from last year to this year.

Condensed Income Statement

Blue Bayou Restaurant

For the Year Ended December 31, 20XX

Sales	$800,000
Cost of sales	225,000
Labor	250,000
Other operating expenses	150,000
Depreciation	50,000
Interest	15,000
Income taxes	20,000
Net income	**$ 90,000**

Figure 5.5 Condensed Income Statement

Statement of Retained Earnings

Blue Bayou Restaurant

For the Year Ended December 31, This Year

Retained Earnings, December 31, Last Year	$300,000
PLUS, Net Income for This Year	+ 90,000
Subtotal	= 390,000
MINUS, Cash Dividends Paid This Year	− 100,000
Retained Earnings, December 31, This Year	= $290,000

Figure 5.6 Statement of Retained Earnings

year's balance sheet are also needed to report the changes in balance sheet accounts from one year to the next. These will show the sources and uses of funds that will affect the cash changes reported on the SCF.

The recommended format for an SCF consists of:

✓ Cash flow from operating activities
✓ Cash flow from investing activities
✓ Cash flow from financing activities
✓ Net changes in cash
✓ Supplementary schedules

Figure 5.7 is an example of the recommended standard format used to prepare an SCF.

Cash Provided (inflow) or Used (outflow) by:	
Operating activities	$XXX.xx
Investing activities	$XXX.xx
Financing activities	$XXX.xx
Net increase (decrease) in cash	$XXX.xx
Cash at beginning of accounting period	$XXX.xx
Cash at end of accounting period	$XXX.xx
Supplementary Schedule of Noncash Investing and Financing Activities	$XXX.xx
Supplementary Disclosure of Cash Flow Information	
Cash paid during the year for:	
Interest	$XXX.xx
Income taxes	$XXX.xx

Figure 5.7 Statement of Cash Flows Format

Cash Flow from Operating Activities

Cash flow from operating activities is the result of all transactions and events that normally make up a business's day-to-day operational activities. These include cash generated from selling products or providing services and income from items such as interest and dividends. Operating activities also include cash payments for items such as food and beverage inventories, payroll, taxes, interest, utilities, and rent. The net amount of cash provided (or used) by operating activities is a key figure on an SCF statement because it shows cash flows over which foodservice operators have the most direct control.

To better understand cash flow from operating activities, consider that foodservice operators have cash inflows from the following operating activities:

✓ Sales of food, beverages, and services
✓ Interest income from money held in bank accounts
✓ Dividend income (if any)
✓ Income from all other non-investment or financing activities (for example, money received to settle a lawsuit or for an insurance settlement)

These operators incur operating activity expenses that cause cash to flow out of their business, including cash outflows for:

✓ Food, beverages, and other items purchased from the operation's vendors for eventual resale to guests
✓ Salaries, wages, and payroll-related taxes for employees
✓ Items such as rent, utilities, insurance, and other related costs of operating the business
✓ Taxes, duties, fines, fees, and penalties imposed by government entities
✓ All other payments not defined as investing or financing activities (for example, money paid to settle a lawsuit or cash contributions to charities)

What Would You Do? 5.1

"I still don't get it," said Ahmed, "how can all the monthly loan repayments we make to the bank for the dining room expansion project we completed last year *not* be considered an operating expense? We make the payments out of the money we get from being able to serve more customers. That's income. Our loan payments should be an expense!"

Ahmed was a new assistant manager at the Causeway Bay restaurant, and he was talking to Nicole, the restaurant's accountant. Nicole was preparing the operation's SCF and was explaining to Ahmed that debt payments, like those to pay off the loan the restaurant used to expand the dining room last year, were not classified as operating expenses when preparing an SCF.

"Well Ahmed," replied Nicole, "It can seem confusing, but think of it this way: assume you owe $1,000 on your credit card, and you have $400 in your checking account and no other savings. Your net worth is -$600. If you write a $100 check and send it to your credit card company, you've now got $900 in debt and $300 in your checking account. Right?"

"I get it now," said Ahmed, "my net worth in cash would still be -$600. And I will still be paying interest until the balance is zero!"

"Right," said Nicole, "your expense occurred when you put the $1,000 on your credit card, not as you pay it off. The interest payments are a whole other matter!"

Assume you were the owner of the Causeway Bay. Why is it important that you know the amount of interest paid on money borrowed for expanding or operating your business as well as the loan amount?

Find Out More

Some minor differences exist between the financial reporting that occurs when (1) applying the generally accepted accounting practices (GAAP) utilized in the United States and (2) reporting that occurs when applying IFRS, the accounting standards developed by the International Accounting Standards Board (IASB) and used in many parts of the world.

One area of difference affects the development of a business's SCF. When using GAAP, interest received or paid is classified as an operating activity. Dividends received are classified as operating activities, and dividends paid are classified as financing activities.

Under IFRS recommendations, cash flows from interest and dividends received or paid must be classified in a consistent manner as either operating, investing, or financing activities. For most businesses, the IFRS recommendation is that interest and dividends paid may be classified as operating or financing cash flows, and interest and dividends received may be classified as operating or investment activities.

To learn more about the entity that establishes accounting standards used worldwide, enter "International Accounting Standards Board (IASB)" in your favorite browser and view the results.

Managers can use two methods when preparing the operating activities portion of an SCF: the direct method and the indirect method. The direct method uses cash receipts from operations and cash disbursements to create the income statement on a cash (not accrual) accounting basis. The indirect method starts with net income as calculated on an accrual basis and then adjusts that number for changes in current assets and current liability accounts.

Although both methods produce identical results, the indirect method is more popular because it more *easily* reconciles the difference between net income and the net cash flow provided by operations. An easier method (when the results are the same) is nearly always a better accounting approach to use than a more difficult method.

The first step in creating an SCF using the indirect method for determining operational cash flows is to develop a summary of cash inflows and outflows resulting from operating activities. As shown in Figure 5.5, managerial accountants need information provided by the income statement including sales, expenses, and net income (profit).

Accountants begin the SCF development process by determining the net income from the income statement, and then they adjust this amount up or down to account for any income statement entries not providing or using cash.

In an accrual accounting system (see Chapter 1), revenue is recorded when it is earned regardless of when it is collected, and expenses are recorded when they are incurred, regardless of when they are paid. Therefore, an accrual-based income statement must be converted to a cash basis to report cash flow from operating activities.

Although there may be several items on the income statement that may need to be adjusted from an accrual basis to a cash basis, the two most common are

✓ Depreciation
✓ Gains or losses from a sale of investments or equipment

Recall that depreciation involves allocating the cost of a fixed asset over its useful life (see Chapter 2). More important, however, depreciation is *subtracted* from the income statement to lower income and, therefore, lower taxes. The portion of assets depreciated each year is considered "tax deductible" because it is subtracted on the income statement before taxes are calculated. Net income is "artificially" lowered by subtracting depreciation. Specifically, no one is writing a check to "depreciation," no cash is actually spent for it, and the cash remains in a business's bank account.

The major noncash items typically expensed in most foodservice operations are depreciation and amortization. Since these expenses are *deducted* on the accrual income statement (but no cash changes hands), they must be *added back* to net income to accurately reflect cash generated from operations when preparing an SCF. Put another way, depreciation and amortization expense can be considered a source of cash, and it is added back to net income (or loss).

To illustrate, consider Thomas Flannigan, who uses an accrual accounting system. This year his income statement showed that his operation produced net income of $70,000. Thomas also showed income statement depreciation in the amount of $10,000 even though this was not paid in cash to anyone.

Therefore, to calculate the overall positive effect of taking his depreciation expense, he must "add cash" back to his net income amount as follows:

Net income before depreciation adjustment	$70,000
Depreciation (from the income statement)	$10,000
Net income adjusted for depreciation	$80,000

The depreciation adjustment works the same way whether a business has produced a positive net income (made a profit) or produced a negative net income (showed a loss).

To illustrate, assume that Thomas's business had lost $30,000 this year, and that, as shown on his income statement, his depreciation expense was again $10,000.

To calculate the positive effect of his depreciation expense, Thomas must again "add cash" back to his net income amount as follows:

Net income prior to depreciation adjustment:	($30,000)
Depreciation (from the income statement)	$10,000
Net income adjusted for depreciation	($20,000)

Adjusting net income for the effect of depreciation is important, but to calculate cash flow from operating activities, net income must also be adjusted for gains or losses from a sale of property and equipment. A gain on a sale of property or equipment occurs when the original cost of the item is lower than the price at which it is later sold. Conversely, a loss on a sale of an investment or equipment occurs when their original cost is higher than the price at which they are later sold.

For example, assume Thomas had $150,000, and he purchased some land to expand his business. Five years later, however, he changed his mind about expansion and sold the land for $200,000. The difference between the original purchase price and the income he received at his selling price is $50,000 and is calculated as follows:

Income from sale	$200,000
Less: Original purchase price	$150,000
Gain from sale	$ 50,000

Recall that losses or gains on the disposition of long-term assets are reported on the income statement. As a result, Thomas will report the gain, but the amount of the gain would be *subtracted* from net income when developing the Cash Flows from Operating Activities portion of his SCF. The reason: The gain was not a part of his normal business operations. The remaining $150,000 cash received would be reported in the Cash Flows from Investments portion of the SCF.

The approach is the same if a loss on a sale is incurred. To illustrate, assume Thomas sold his piece of land for only $100,000. In that case the difference between his original purchase price and the income he received reflects a loss of $50,000 and is calculated as follows:

Income from sale	$100,000
Less: Original purchase price	$150,000
Loss from sale	($ 50,000)

In this scenario, $50,000 must be added back to the net income shown on Thomas's income statement when developing the Cash Flow from Operating

Activities portion of his SCF since the loss was not a part of his normal business operations.

The remaining adjustments to net income when calculating Cash Flows from Operating Activities come from changes in Deferred Taxes accounts and from the sources and uses of funds calculated from the balance sheet. It is important to remember that sources of funds are shown as a positive number on the SCF, and uses of funds are shown as a negative number.

In general, the sources and uses of funds needed to prepare the Cash Flows from Operating Activities portion of an SCF come from changes in a business's current assets and current liabilities accounts (from the balance sheet) since these typically represent current operational activities.

Additional adjustments are then made for depreciation, amortization, and gains or losses from the sale of property and equipment, and from any changes in deferred taxes accounts. Note: Any changes in the value of short-term investments (marketable securities) reported on the income statement will be accounted for in the Cash Flows from Investing Activities portion of the SCF. Any interest paid on borrowed money reported on the income statement is adjusted in the Cash Flows from Financing Activities portion of the SCF.

Cash Flow from Investing Activities

An investment is the acquisition of an asset to increase future financial return or benefits. Cash flow from investing activities summarizes this part of a business's action. Investing activities include transactions and events involving the purchase and sale of marketable securities, investments, land, buildings, equipment, and other assets not generally purchased for resale to customers.

For example, if an operator elected to purchase land adjacent to the business to expand the size of the parking lot, the cash inflows and outflows associated with that purchase would be accounted for as an investment activity. The reason: The operator is buying an item (land) not intended to be re-sold to guests but rather to improve the business' long-term financial performance. While the purchase may be critical to the continued success of the business, the purchase is not recorded in cash flow from operating activities.

Investing activities are not classified as operating activities. The reason: In a foodservice operation investing activities are considered to have less of a direct relationship to the central, ongoing operation of the business than will the sale of food and beverage items.

Examples of cash inflows and outflows related to the investing activities of a business include:

Cash inflows:

✓ The sale of short-term investments (marketable securities) owned by the business
✓ The sale of long-term investments
✓ The sale of property, buildings, furnishings, equipment, and other assets
✓ Income from all other non-operating or financing activities

Cash outflows:

✓ The purchase of marketable securities (short-term investments)
✓ The purchase of long-term investments
✓ The purchase of property, buildings, furnishings, equipment, and other assets
✓ All other payments not defined as operating or financing activities

The cash flow from investing activities comes from the sources and uses of funds shown on the balance sheet. For investment activities operators, should remember that sources of funds are shown as a positive number on an SCF, and uses of funds are shown as a negative number.

Cash Flow from Financing Activities

The third of the three cash inflow and outflow activity summaries needed to create a complete SCF relates to the financing activities of a business. Cash flow from financing activities refers to a variety business activities including:

✓ Obtaining resources (funds) from the owners of a business (i.e., by selling company stocks)
✓ Providing owners with a return of their original investment amount (i.e., by the payment of dividends)
✓ Borrowing money
✓ Repaying borrowed money

When a business corporation sells portions of ownership in itself by issuing stock, declaring a stock dividend (stockholders' payments based on the number of stock shares owned), re-purchasing its own stock, and borrowing or paying back money, it is involved in a financing activity.

Remember that, even though loan repayments are a financing activity, interest paid and received are classified as operating activities (and are reported on the income statement). For example,

Key Term

Principal (loan): The amount a borrower agrees to pay a lender when a loan becomes due, not including interest.

if an operator wanted to purchase land for a parking lot by taking out a loan, cash payments made to reduce the **principal** (the amount borrowed) of the loan would be considered cash flow related to a financing activity, while any interest paid to secure the loan would be considered an operating expense. Loans, notes, and mortgages are examples of financing activities that affect cash flows.

Examples of cash inflows and outflows related to the financing activities of a business include:

Cash inflow from:

✓ Funds obtained from short-term borrowing
✓ Funds obtained from long-term borrowing
✓ Proceeds from the issuance (sale) of stock
✓ Income from all other non-operating or investment activities

Cash outflow from:

✓ Repayment of loans (short-term and long-term loans)
✓ Re-purchase of issued stock
✓ Dividend payments to stockholders
✓ All other payments not defined as operating or investing activities

The cash flow from financing activities comes from the sources and uses of funds calculated from an operation's balance sheets. For financing activities, operators must remember that sources of funds are shown as a positive number on an SCF, and uses of funds are shown as a negative number.

In general, the sources and uses of funds needed to determine cash flow from financing activities will come from long-term debt and equity. The exception to this is notes payable (short-term debt), a current liability that belongs in financing activities. Also, dividends paid must be recorded here because they are a cash outflow from net income.

Common additions and subtractions to the SCF are shown in Figure 5.8. For many operators, these are easier to remember when thinking in terms of the balance sheet. With the exceptions noted, operating activities are developed using current assets and current liabilities, investing activities are developed using long-term asset accounts, and financing activities are developed using long-term debt and owners' equity accounts.

Net Changes in Cash

Net changes in cash represent all cash inflows minus all cash outflows from operating, investing, and financing activities. It is, in effect, the total net change when combining all three

Key Term

Net changes in cash: The total increase or decrease of cash and cash equivalent balances within a specified accounting period factoring in the net changes in cash for operating, investing, and financing activities.

Operating activities
Net income
+/− Depreciation
+/− Losses/gains from the sale of assets
+/− Current assets (except marketable securities)
+/− Current liabilities (except notes payable)
Investing activities
+/− Investments and marketable securities
+/− Investments
+/− Property and equipment
+/− Other assets
Financing activities
+/− Notes payable
+/− Long-term debt
+/− Common stocks and paid in capital
+/− Dividends paid

Figure 5.8 Common Additions and Subtractions to the Statement of Cash Flows

activities. This net change in cash must equal the difference between the cash account at the beginning of the accounting period and the cash account at the end of the accounting period. If the net change does not equal the difference between the beginning and ending cash balances, the SCF was not prepared properly.

To illustrate net change in cash, assume a foodservice operator has less money in her cash account at the end of an accounting period than she did at the beginning of the period. She carefully prepares her SCF with these results:

Cash Flow from Operating Activities		($150,000)
Cash Flow from Investing Activities	Plus	$100,000
Cash Flow from Financing Activities	Plus	$ 25,000
Net decrease in cash	Equals	($ 25,000)
Cash at the beginning of the period		$625,000
Cash at the end of the period	Minus	$600,000
Net decrease in cash account	Equals	$ 25,000

Notice that the operator's net *decrease* in cash of $25,000 equals the difference between her cash at the beginning and the ending of the period: $625,000 − $600,000 = $25,000. In this example, the operator has shown why she has $25,000 *less* in her cash account at the end of the accounting period than she had at the beginning of the period.

Now assume that the same operator had calculated the following results:

Cash Flow from Operating Activities		$ 150,000
Cash Flow from Investing Activities	Plus	$(100,000)
Cash Flow from Financing Activities	Plus	$ (25,000)
Net increase in cash	Equals	$ 25,000
Cash at the end of the period		$ 675,000
Cash at the beginning of the period	Minus	$ 650,000
Net increase in cash account	Equals	$ 25,000

In this second scenario, her net *increase* in cash of $25,000 equals the difference between her cash account at the end of the period and her cash account at the beginning of the period: $675,000 − $650,000 = $25,000. In this example, the operator has $25,000 *more* in her cash account at the end of the accounting period than at the beginning of the period.

It is important to recognize that, whether cash increases or decreases, if the net change in cash is *not* equal to the change in the cash account, an accountant must review their work to find the error (and there will be one!).

Supplementary Schedules

Supplementary schedules to the SCF are used when accountants want to communicate additional information such as the reporting of noncash investing and financing activities and cash that was paid out for interest and income taxes.

Businesses can, of course, increase their assets and/or decrease their liabilities without utilizing cash. Consider, for example, the owners of a piece of land valued at $100,000. Mike Connell, the owner of a chain of sub shops wants to buy the land to open a new unit. Mike is told by the land's owners that they would be willing to exchange the piece of land for shares in his company's stock.

If Mike were to accept the offer valued at $100,000, his balance sheet would change (with an increase in the long-term asset portion of the balance sheet titled "Land" and a corresponding increase in the Owner's Equity portion of the balance sheet). However, the cash position of his business would not have changed because this would have been a **noncash transaction**.

Significant noncash investing and financing transactions undertaken by a company should be reported in a Supplementary Schedule of Noncash Investing and Finance Activities that is attached as a supplement to the SCF.

Key Term

Noncash transaction: Investing- and financing-related transactions that do not involve the use of cash or a cash equivalent.

Since noncash investing and finance activities don't involve cash, they are not reported on the SCF. However, these transactions still provide very useful information about an organization's investing and financing activities. Any significant noncash investing and financing activities must be reported in a supplementary schedule shown at the bottom of the SCF or in the Notes to the Financial Statements section.

The requirement to disclose significant noncash investing and financing activities is consistent with the full disclosure principle of GAAP; any information that would make a difference to the readers of a financial statement should be provided to them.

Below is a simplified version of a noncash investing and financing activities schedule that would explain the exchange of Mike's stock for title to the land he plans to use for his new coffee shop.

Supplementary Schedule of Noncash Investing and Financing Activities

Stock exchange for land (for new coffee shop)	<u>$100,000</u>

Also included in the SCF (and required by the FASB) is the Supplementary Disclosure of Cash Flow Information, which reports cash paid during the year for interest and income taxes.

When properly prepared, an SCF will include:

✓ A summary of cash inflows and outflows resulting from operating activities
✓ A summary of cash inflows and outflows resulting from investing activities
✓ A summary of cash inflows and outflows resulting from financing activities
✓ Net changes in cash from the beginning to the end of the accounting period
✓ A supplementary schedule of noncash investing and financing activities (if applicable)
✓ A supplementary disclosure of cash flow information (for interest and income taxes)

To illustrate each of the above, a sample of a fully completed SCF for Rhonda's Bagel Shop is presented in Figure 5.9.

Technology at Work

In most cases, those who operate or own a foodservice business will rely on a professional accountant to create their statement of cash flows (SCF).

However, for those operators who wish to create their own SCF, there are several large and small software companies that can help, and SCF production software is relatively inexpensive and easy to use.

To review some of the SCF product offerings appropriate for foodservice operators, enter "statement of cash flows software for restaurants," in your favorite search engine and review the results.

Statement of Cash Flows

Rhonda's Bagel Shop

For the Year Ended December 31, 20XX

Net Cash Flow from Operating Activities		
Net Income		$74,000
Adjustments to reconcile net income to net cash flows		
from operating activities		
Depreciation	$50,000	
Gain on sale of investments	−10,000	
Decrease in accounts receivable	4,000	
Increase in inventory	−2,000	
Increase in accounts payable	500	
Increase in accrued payroll	500	
Decrease in income taxes payable	−1,000	42,000
Net Cash Flow from Operating Activities		116,000
<u>Investing Activities</u>		
Sale of investments	$5,000	
Purchase of investments	−15,000	
Purchase of equipment	−20,000	
Net Cash Flow Used in Investing Activities		−30,000
<u>Financing Activities</u>		
Payment of long-term debt	−26,000	
Dividends paid to stockholders	−30,000	
Net Cash Flow Used in Financing Activities		−56,000
Net Increase in Cash in 20XX		30,000
Cash at Beginning of 20XX		20,000
Cash at End of 20XX		$50,000
Supplementary Disclosure of Cash Flow Information		
Cash paid during the year for:		
Interest		$34,000
Income taxes		$28,000

Figure 5.9   Sample Completed Statement of Cash Flows (SCF)

Statement of Cash Flows Analysis

The SCF is a financial document that is not analyzed nearly as much as the income statement or balance sheet. This is partially because the SCF is relatively new to the managerial accounting world (it has only been required by the FASB since 1988). Also, many managers' jobs are more concerned with operations (income statement) and the effective use of assets (balance sheet) than cash management because the management of cash flows is most often the job of an operation's owner, **Chief Financial Officer (CFO)**, or accountant.

Despite the lesser analysis of the SCF, the ability of a business to generate cash from its core business operations is an important indicator of its financial health and the risk that investors will take when they invest in it.

As a result, one good method of analyzing the SCF is to compare operating, investing, and financing activities from last year to this year. The dollar amount of changes from previously experienced levels on the balance sheet is an indication of whether an operation's cash position is improving, declining, or staying the same.

While cash flows in a business are certainly no absolute guarantee of future cash flows, it is a good indication of how well company managers are generating cash flows. In addition to knowing how much (and when) cash is generated, investors also want to know how management is using the cash it accumulates. For example, assume that an operator generates $100,000 in positive cash flows in the company they manage. The owners of the company may, for example, elect to use all or part of this cash to:

✓ Pay dividends to investors
✓ Retire some of the company's long-term debt
✓ Buy additional land for new development
✓ Give managers bonuses
✓ Give employees raises
✓ Invest in new equipment
✓ Leave the money in the company's bank account

Therefore, it is not merely how much cash is generated, but the ultimate use of the cash that should also be considered after preparing an SCF.

Management's ultimate use of cash is often equally as important as how much cash is available to spend and, for many investors and managers, a business's **free cash flow** is an important measure of its economic health.

Free cash flow is considered a good measure of a company's ability to pay its debts, ensure its growth, and pay (if applicable) dividends to its investors. To illustrate the importance of free cash flow, consider two different operations.

The first operation had a cash income of $50,000 last year but has loan payments of $40,000 due in the coming year, resulting in $10,000 in free cash flow ($50,000 − $40,000 = $10,000).

The second operation had a cash income of $40,000 last year and has loan payments of $10,000 due in the coming year, resulting in $30,000 in free cash flow ($40,000 − $10,000 = $30,000).

Despite its lower gross cash flow (income), the second operation has a better free cash flow. In business terms, the second operation would be better able to withstand a sales downturn or similar short-term financial difficulty than would the first operation.

The formula managerial accountants use to calculate free cash flow for their companies is:

	Net cash provided from operating activities
Minus:	Cash used to acquire property and equipment
Equals	Free cash flow

A company with a large and positive free cash flow can grow and invest its excess cash in its own expansion, alternative investments, or by returning money to the company's owners. Managers of a business can influence their free cash flow by taking longer to pay their bills (preserving cash), aggressively collecting their accounts receivable (which accelerates the receipt of cash), or not buying needed inventory (which preserves cash).

Savvy operators know that negative free cash flow is not always a bad thing. If, for example, a company is making large investments in property and equipment, and these investments earn a high return on investment (ROI) in the near future, the strategy has the potential to be sound in the long run. If, however, a company has a consistently negative free cash flow, it will likely need to supplement its cash from other sources, and this would involve borrowing funds or seeking additional investors. Since external funding for a company is rarely unlimited, a company with a consistent (and significant sized) negative free cash flow simply must turn that situation around or face the ultimate closing of the business.

The income statement, balance sheet, and SCF are important financial documents all operators must understand to assess their profitability. Since profits are the dollars remaining after an operation subtracts its costs from its revenue, operators must also understand the "cost" of operating their businesses. The concept of cost and the proper accounting for costs is the topic of the next chapter.

What Would You Do? 5.2

Benji Metzger had just received his Statement of Cash Flows from his accountant Lyla.

She had e-mailed the document with a short note congratulating him for another successful year as the owner of Benji's Bistro, a 70-seat restaurant that featured healthy foods served in a casual atmosphere.

According to the SCF document Lyla had sent him, his free cash flow for the year was just over $100,000, and that was after paying the monthly salary that Benji took for himself as the manager of the business.

As he reviewed the statement, Benji wondered if he should take all the $100,000 and invest it in the 401(k) retirement program he had started for his wife and him. He liked that idea, but he also knew that his dining area had not been renovated for the past six years, and it was beginning to show its age. A remodel of that area was going to be needed soon, and it would be costly.

Assume you were Benji. What could be some reasons you would want to take the excess cash from the business and invest it in your retirement program? What would be some reasons it might make good sense to re-invest it in your own business to upgrade the dining room area?

Key Terms

Cash	Going concern	Principal (loan)
Bankruptcy	Cash equivalent	Net changes in cash
Cash management	Sources and uses of funds	Noncash transaction
Statement of Cash Flows (SCF)	Condensed income statement	Chief Financial Officer (CFO)
Consolidated income statement	Statement of retained earnings	Free cash flow

Operator's 10-Point Tactics for Success Checklist

Evaluate your need for, and the current status of, each of the following operational tactics. For those tactics you think are important, but not yet in place, develop an action plan for its implementation including who will be responsible for the tactic's completion and the target date by which it should be completed.

Tactic	Don't Agree (Not Done)	Agree (Done)	Agree (Not Done)	If Not Done Who Is Responsible?	Target Completion Date
1) Operator understands the important role of cash and cash management in the operation of a successful business.	___	___	___		
2) Operator recognizes the purpose of producing a Statement of Cash Flows (SCF).	___	___	___		
3) Operator is aware that sources of funds represent a business's cash inflows.	___	___	___		
4) Operator is aware that uses of funds represent a business's cash outflows.	___	___	___		
5) Operator can calculate the amount of cash flow resulting from their operating activities.	___	___	___		
6) Operator can calculate the amount of cash flow resulting from their investing activities.	___	___	___		
7) Operator can calculate the amount of cash flow resulting from their financing activities.	___	___	___		

				If Not Done	
Tactic	**Don't Agree (Not Done)**	**Agree (Done)**	**Agree (Not Done)**	**Who Is Responsible?**	**Target Completion Date**
8) Operator recognizes that "Net Change in Cash" is the sum of net changes in cash resulting from all operating, investing, and financing activities.	——	——	——		
9) Operator appreciates the value of supplemental schedules prepared for the purpose of providing additional information to a Statement of Cash Flows (SCF).	——	——	——		
10) Operator understands the importance to success of "Free Cash Flow;" as well as how free cash flow is calculated.	——	——	——		

6

Understanding Costs and Break-even Analysis

> **What You Will Learn**
> 1) The Importance of Understanding Business Costs
> 2) How to Identify the Different Types of Costs
> 3) How to Perform a Break-even Analysis When Costs Are Known

> **Operator's Brief**
>
> A cost (expense) in a foodservice operation is often defined as "time or resources expended." A simpler explanation of a business cost is the amount of money needed to do, buy, or make something. These activities describe foodservice operators because they do, buy, or make things that are then sold to their guests.
>
> There are several types of costs, and in this chapter, you will learn about different types of foodservice costs managerial accountants consider when they analyze the total costs incurred by an operation. Important costs include fixed and variable; mixed; step; direct and indirect (overhead); controllable and non-controllable; and other types of costs described in the chapter.
>
> Perhaps the most important cost concept for you to understand is that, in nearly all cases, as the number of items sold to your guests increases, the total costs associated with providing those items typically increase as well. As sales increase, the total dollars expended for costs increase; but as a percentage of revenue, your costs will decline, and your profits will increase.
>
> Analyzing the important relationship among the cost, number of items sold (volume of sales), and profitability is called cost/volume/profit analysis. The more popular term for this concept is "break-even" analysis.
>
> In this chapter, you will learn how to use information from your business to perform a break-even analysis. You will discover that it is important to understand the break-even concept so you can use it to help better manage your foodservice operation.

CHAPTER OUTLINE

The Importance of Understanding Costs
Types of Costs
    Fixed and Variable Costs
    Mixed Costs
    Step Costs
    Direct and Indirect (Overhead) Costs
    Controllable and Non-controllable Costs
    Other Types of Costs
Break-even Analysis
    Computation of Break-even Point
    Minimum Sales Point

The Importance of Understanding Costs

"Cost" is a word commonly used in the foodservice industry. As a result, operators can, among other things:

✓ Control costs
✓ Determine costs
✓ Cover costs
✓ Cut costs
✓ Measure rising costs
✓ Eliminate costs
✓ Estimate costs
✓ Budget for costs
✓ Forecast costs

Given all the possible approaches to examine costs, perhaps the easiest way to understand them is to consider their impact on business profits.

Recall that the basic profit formula was presented in Chapter 3 as:

$$Revenue - Expense = Profit$$

In this formula, cost and expense mean the same thing and, in common usage, they are often used interchangeably. Expressed in alternative phrasing and using algebra, the basic profit formula becomes:

$$Revenue = Profit + Costs$$

As the formula shows, at any specific level of revenue, the *lower* a business's costs, the greater are its profits. This is why understanding and controlling costs are so important in the successful operation of a foodservice business.

Cost (expense) is the reduction of a business asset, generally for the ultimate purpose of increasing an operation's revenues. However, not all costs are the same, and it is easy to see why all costs cannot be viewed in a similar manner.

Consider the foodservice operator who has a monthly mortgage payment that must be made regardless of the amount of revenue the operator's business generates each month.

If the operator knew in advance the guest check average (see Chapter 1) that would be achieved in a specific month, it would be easy to compute the number of total guests required to meet the monthly mortgage obligation. However, people do not come to a foodservice operation so its owner can pay the mortgage. Instead, they come to enjoy the products and services the operation offers. As a result, as more people visit the operation, the number of meals provided to them also is greater. The operation's "cost of mortgage" remains the same regardless of the number of guests served, but the cost of providing meals varies.

Both the mortgage and meals cost represent business costs, but they are very different types of cost. In fact, there are a variety of useful ways foodservice operators and managerial accountants can view costs and then better understand and operate a business.

Types of Costs

Since not all costs are the same, professional cost accountants have identified several ways to classify business costs. Among them, the most important to understand are:

✓ Fixed and variable costs
✓ Mixed costs
✓ Step costs
✓ Direct and indirect (overhead) costs
✓ Controllable and non-controllable costs
✓ Other types of costs:
  • Joint costs
  • Incremental costs
  • Standard costs
  • Sunk costs
  • Opportunity costs

Before examining the various types of costs, operators should recognize that not all business costs can be objectively measured. In fact, sometimes cost management can be subjective.

To illustrate, consider that identifying a cost can be as simple as reviewing an invoice for purchased meats or produce. For example, the operator of a small sub shop orders and receives several hundred dollars' worth of produce for the next few days of production. They know exactly what the items "cost" because the vendor's invoice states the amount.

In other cases, however, "cost" is more subjective. Consider the salary of an operator at a small business concerned about the quality of some produce just delivered. They contact the applicable vendor to discuss replacement of items in the current order. The cost of preparing payments and discussing them is an activity performed by many foodservice operators for which a clear-cut cost is not easily assigned. Some customers (purchasers) have invoice questions and others will not. How can the real cost of "preparing bills and talking with vendors about invoice questions" be assigned?

Cost accountants confronted with these issues can assign each foodservice employee's time to different activities performed within a business. Surveys might be used by workers themselves who assign their time to different activities such as food preparation, guest service, and clean-up. An accountant can then determine the total cost spent on each activity by summing the percentage of each worker's time and compensation spent on an activity. This process is called **activity-based costing**. The goal is to assign objective costs to subjective items including payment for various types of labor and even more subjective management tasks involved with planning, organizing, directing, and controlling a foodservice operation.

Key Term

Activity-based costing:
A method of assigning overhead and indirect costs such as salaries and utilities to specific products and services.

It is easy to understand how allocating costs to various areas of a business can permit managers to total up and then examine various costs to make good decisions. If an operator believes too little money is spent in one area (e.g., dining room servers), additional costs can be allocated to that area. If too much is being spent (e.g., on utilities), actions can be taken to correct and control the costs in that area.

Using activity-based costing to examine expenses and better manage a business is one example of how more fully understanding costs can help a foodservice operator make better decisions to manage a successful business.

Fixed and Variable Costs

Some foodservice operations' costs stay the same each month. For example, if an owner elected to buy a building for a restaurant and cocktail lounge, the owner's mortgage payment is generally set up so the same dollar amount is paid each month. Likewise, if the owner wanted to provide overhead music in the restaurant, the amount paid for the rights to play the music will also be the same each month for the life of the contract signed with the music provider selected.

As a final example, if the owner elected to lease, rather than purchase, a new dishwasher for the business, the monthly amount of the lease payment will likely be at a specified dollar value. In each of these examples, the cost involved is the same (fixed) each month. For that reason, these types of costs are referred to as fixed costs (see Chapter 3).

A fixed cost is one that remains constant despite increases or decreases in sales volume (the number of guests served or average guest check size). Other typical examples of fixed costs include payments for insurance policies, property taxes, and management salaries.

The relationship between sales volume and fixed costs is shown in Figure 6.1, where the cost in dollars is displayed on the *y* axis, and sales volume is shown on the *x* axis. Note that the cost is the same regardless of sales volume.

In some cases, the amount an operator pays for an expense will not be fixed; instead, it will vary based on business success. For example, in a nightclub, the expenses incurred for paper cocktail napkins will increase as the number of guests served increases, and napkin costs will decrease as the number of guests served decreases.

Similarly, in a steakhouse restaurant, each steak sold requires the operator to purchase another steak to maintain the appropriate inventory. In this example,

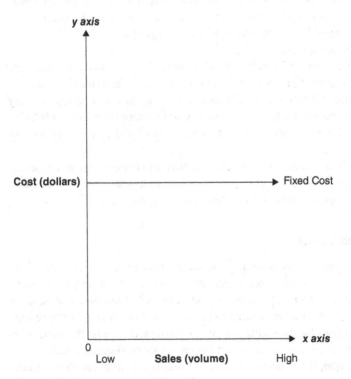

Figure 6.1 Sales Volume and Fixed Costs

steak costs increase when steak sales increase and decrease when steak sales decrease. Stated differently, steak costs *vary* with volume.

A variable cost (see Chapter 3) increases as sales volume increases and decreases as sales volume decreases. The relationship between sales volume and variable cost is shown in Figure 6.2 in which the cost in dollars is displayed on the y axis, and sales volume is shown on the x axis.

To illustrate the calculation of variable costs, consider Brenda's Bountiful Burgers, a midsize, freestanding restaurant. Brenda features upscale gourmet burgers, and ingredient costs are $4.00 to make a gourmet burger. If 50 guests order burgers, her total variable food cost for burgers can be calculated as:

Variable Cost per Guest (VC/Guest) × Number of Guests = Total Variable Cost

Or

$$\$4.00 \times 50 = \$200$$

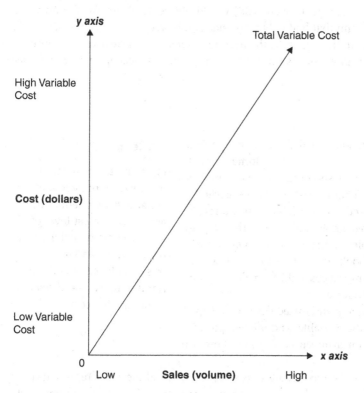

Figure 6.2 Sales Volume and Variable Costs

If the total variable cost and the number of guests is known, variable cost per guest (VC/Guest) can be determined. Using basic algebra, a variation of the total variable cost formula can be calculated as:

$$\frac{\text{Total Variable Cost}}{\text{Number of Guests}} = \text{VC/Guest}$$

Or

$$\frac{\$200}{50} = \$4.00$$

Experienced foodservice operators want to decrease their fixed costs to their lowest practical levels while still satisfying the needs of the business and its customers. Those same operators know that *increases* in variable costs are usually good signs of business activity.

The owner of a steakhouse restaurant that incurred "zero" steak expense last week (because zero steaks were sold) would not be as pleased as the owner of another restaurant that had to buy 500 steaks this week to replace the 500 steaks sold last week. In this example, the need to purchase extra steaks and incur extra variable costs simply means the operation sold more steaks because it increased its sales!

Mixed Costs

It is clear that some foodservice costs are fixed and that some vary with sales volume (variable costs). Still other costs contain a mixture of fixed and variable characteristics. These costs are called semi-fixed, semi-variable, or **mixed costs**. All these names imply that costs of this type are neither completely fixed nor completely variable. In this book, these types of cost will be referred to as mixed costs: the term best explaining their true nature.

To better understand mixed costs, it is helpful to see how the variable and mixed portions are depicted on a mixed cost graph shown in Figure 6.3.

Key Term

Mixed cost: A cost composed of a mixture of fixed and variable components. Costs are fixed for a set level of sales volume or activity and then become variable after this level is exceeded. Also referred to as a semi-fixed or semi-variable cost.

In Figure 6.3, the *x* axis represents sales volume, and the *y* axis represents costs (in dollars). The total mixed cost line is a combination of fixed and variable costs. The mixed cost line starts at the point where fixed costs meet the *y* axis because

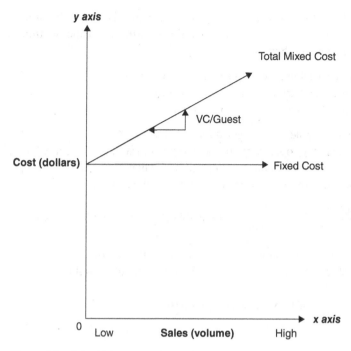

Figure 6.3 Sales Volume and Mixed Costs

fixed costs are the same regardless of sales volume, and they must be paid even if no sales occur. The variable cost line, then, sits on top of the fixed cost line, and each guest served generates a portion of the VC/guest. Therefore, the total mixed cost line includes a combination of the fixed costs and variable costs needed to generate total sales.

In a foodservice operation, labor is most often considered a mixed cost because the operation typically employs salaried employees (a fixed cost) and hourly employees (a variable cost).

Separating Mixed Costs into Variable and Fixed Components
Mixed costs include both variable and fixed cost components. A mixed cost can be divided into its fixed and variable components. Then, managers can effectively control the variable cost portion, and this variable cost is the most often controllable in the short term.

Several procedures can split mixed costs into fixed and variable components. One good way to do so is by use of the "high/low" method. This method is easy to use and gives a good estimate of the variable and fixed components of a mixed cost.

To illustrate the high/low method, consider the Golden Dragon Restaurant. The restaurant's costs, including the number of guests served, from its income statements for October, November, and December are shown in Figure 6.4.

To demonstrate how the high/low method separates a mixed cost into its variable and fixed components, consider the operation's *Marketing* expense from Figure 6.4. The high/low method uses three steps:

Step 1: Determine variable cost per guest for the mixed cost.
Step 2: Determine total variable costs for the mixed cost.
Step 3: Determine the fixed costs portion of the mixed cost.

Step 1: Determine variable cost per guest for the mixed cost.

To complete this step, operators choose a high-volume month and a low-volume month that represents *normal* operations. Then, using the following formula, they separate out variable cost per guest for the mixed cost:

$$\frac{\text{High Cost} - \text{Low Cost}}{\text{High \# of Guests} - \text{Low \# of Guests}} = \text{Variable Cost per Guest (VC/Guest)}$$

In the Golden Dragon example, and using December as the high-volume month and October as the low-volume month, the calculation would be:

$$\frac{\$3,462 - \$2,912}{21,000 - 10,000} = \$0.05 \text{ variable cost per guest}$$

	October	November	December
Number of Guests	10,000	17,000	21,000
Cost of Food	$ 35,000	$ 59,500	$ 73,500
Cost of Beverage	4,000	6,800	8,400
Salaries and Wages	23,960	27,460	29,460
Employee Benefits	5,125	5,265	5,345
Direct Operating Expenses	6,056	8,156	9,356
Music and Entertainment	1,070	1,070	1,070
Marketing	2,912	3,262	3,462
Utility Services	4,077	5,477	6,277
Repairs and Maintenance	1,630	1,840	1,960
Administrative and General	5,570	5,570	5,570
Occupancy	10,000	10,000	10,000
Depreciation	3,400	3,400	3,400
Interest	7,200	7,200	7,200
Total Costs	**$110,000**	**$145,000**	**$165,000**

Figure 6.4   Golden Dragon Restaurant Costs for October, November, and December

Step 2: Determine total variable costs for the mixed cost.

Multiply the variable cost per guest from Step 1 by either the high or low volume (number of guests).

Using October guest counts, the calculation would be:

VC/Guest × Number of Guests = Total Variable Cost

Or

$0.05 × 10,000 (October guests) = $500

Step 3: Determine the fixed costs portion of the mixed cost.

Subtract total variable cost from the mixed cost at the high-volume or low-volume level chosen in Step 2 to determine the fixed cost portion as follows (Note: Low volume at 10,000 (October) guests was chosen in this example.):

Mixed Cost − Total Variable Cost = Fixed Cost

Or

$2,912 − $500 = $2,412

The Golden Dragon's mixed marketing expense can be shown with its variable and fixed components as:

Fixed Cost + Variable Cost = Total Mixed Cost

Or

Fixed Cost + (Variable Cost per Guest × Number of Guests) = Total Mixed Cost

In this example,

At 10,000 guests served:

$2,412 + ($0.05 × 10,000) = $2,912

After utilizing the high/low method for separating mixed costs into variable and fixed components, operators next separate *all* costs into their variable and fixed components. To determine variable costs and fixed cost components at the Golden Dragon, its operator takes several steps:

Step 1: Identify all costs as being variable, fixed, or mixed.

True variable costs at the Golden Dragon will be food costs and beverage costs because they vary directly with the number of guests served.

True fixed costs are music and entertainment, administrative and general, occupancy, depreciation, and interest. These are easy to identify since the costs are the same every month regardless of the number of guests served.

The operation's mixed costs have characteristics of both variable and fixed costs, and they are salaries and wages, employee benefits, direct operating expenses, marketing, utility services, and repairs and maintenance.

Step 2: Determine variable cost per guest for *each* variable cost.

Using the Golden Dragon's food cost as an example, variable cost per guest is $35,000 food cost/10,000 guests = $3.50.

Step 3: Determine *each* fixed cost.

The best way for operators to identify fixed costs is by recognizing those costs that are the same each month.

Step 4: Determine the variable cost and fixed cost portions of *each* mixed cost.

This is done using the high/low method to separate mixed costs into their variable and fixed components as was illustrated previously with the Golden Dragon's marketing expense.

Figure 6.5 shows the variable cost per guest and fixed costs for the Golden Dragon Restaurant.

	October	November	December	Variable Cost per Guest	Fixed Costs
Number of Guests	10,000	17,000	21,000		
Cost of Food	35,000	59,500	73,500	3.50	0
Cost of Beverage	4,000	6,800	8,400	0.40	0
Salaries and Wages	23,960	27,460	29,460	0.50	18,960
Employee Benefits	5,125	5,265	5,345	0.02	4,925
Direct Operating Expenses	6,056	8,156	9,356	0.30	3,056
Music and Entertainment	1,070	1,070	1,070	0	1,070
Marketing	2,912	3,262	3,462	0.05	2,412
Utility Services	4,077	5,477	6,277	0.20	2,077
Repairs and Maintenance	1,630	1,840	1,960	0.03	1,330
Administrative and General	5,570	5,570	5,570	0	5,570
Occupancy	10,000	10,000	10,000	0	10,000
Depreciation	3,400	3,400	3,400	0	3,400
Interest	7,200	7,200	7,200	0	7,200
Total Costs	110,000	145,000	165,000	$5.00	$60,000

Figure 6.5 Golden Dragon Restaurant Variable Costs per Guest and Fixed Costs

As shown in Figure 6.5, the Golden Dragon's total variable cost per guest is $5.00 and its total fixed costs are $60,000. The operation's total costs for October ($110,000), November ($145,000), and December ($165,000) include variable, fixed, and mixed costs.

It is important to recognize that *total* costs are mixed costs, and they can be treated as such. Therefore, by substituting Total Cost for Total Mixed Cost in the Total Mixed Cost formula, the Golden Dragon's Total Costs can be calculated as:

Fixed Costs + Variable Costs = Total Costs

Or

Fixed Costs + (Variable Cost per Guest × Number of Guests) = Total Costs

Or

At 10,000 guests served:
$60,000 + ($5.00 × 10,000) = $110,000

At 17,000 guests served:
$60,000 + ($5.00 × 17,000) = $145,000

At 21,000 guests served:
$60,000 + ($5.00 × 21,000) = $165,000

Total fixed costs and total variable costs per guest are the same for all levels of number of guests served. The reason: The total cost equation represents a straight line as shown in Figure 6.6.

Basic algebra states that the equation for a line is:

$y = a + bx$

The equation for a line applies to the Total Cost line, where a is the y intercept (fixed costs), b is the slope of the line (VC/Guest), x is the independent variable (Number of Guests or Sales Volume), and y is the dependent variable (Total Cost). The *total cost equation* for the Golden Dragon can be summarized as follows:

Total Costs = Fixed Costs + (Variable Cost per Guest × Number of Guests)

Or

At any number of guests served at the Golden Dragon:

Total Costs = $60,000 + ($5.00 × Number of Guests)

Assuming that in a *normal* month variable costs per guest and fixed costs remain the same for the Golden Dragon, total costs in any month can be estimated by using the total cost equation. All its operator must do is insert the anticipated number of guests into the equation to estimate total costs for any month. For

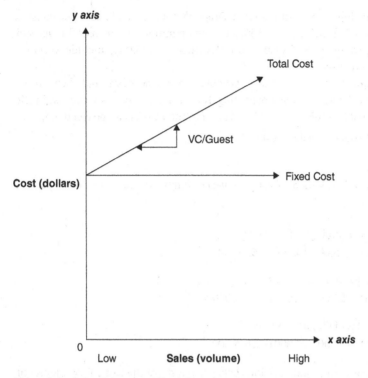

Figure 6.6 Sales Volume and Total Costs

example, if the operator expects that, in June, the restaurant will serve 18,000 guests, its total costs for June can be estimated as:

Total Costs = Fixed Costs + (Variable Cost per Guest × Number of Guests)

Or

At 18,000 *guests served:*

$$\$150{,}000 = \$60{,}000 + (\$5.00 \times 18{,}000)$$

Effective operators know they should not categorize fixed, variable, or mixed costs in terms of being either "good" or "bad." Some costs by their nature relate to sales volume and others do not. It is wise to remember that the goal of management is not to reduce but to *increase* total variable costs in direct relation to increases in total sales volume.

Costs are required to properly service guests, and the more guests served, the greater will be the operation's variable costs. If the total cost of servicing guests is less than the amount received from them, expanding the number of guests will result in increased costs, but it will result in increased profits as well.

Step Costs

Foodservice operators are in the hospitality industry and, in addition to considering their costs as fixed, variable, or mixed, it is also helpful to understand **step costs**. A step cost is a cost that increases as a range of activity increases or as a capacity limit is reached. Instead of increasing in a linear fashion like variable costs (see Figure 6.2), step cost increases look more like a staircase (hence the name "step" costs).

Key Term

Step costs: Costs that increase in a non-linear fashion as activity or volume increases.

It is easy to understand step costs. Consider the foodservice operator who determines one well-trained server can effectively provide excellent service for a range of 1–30 of the operation's guests. If, however, 40 guests are anticipated, a second server must be scheduled. However, the operator does not need a "full" server, and one-third of a server would be sufficient because only 10 additional guests are anticipated.

As shown in Figure 6.7, however, each additional server added increases the operation's costs in a non-linear (step-like) fashion.

A variety of costs can be considered step costs. One of the most important is the cost of managers and supervisors. For example, one banquet supervisor may be able to direct the work activities of eight servers. When nine or more servers are scheduled to work, additional banquet supervisors must be added, and these additional staff members will increase the operation's payroll in a step-like fashion.

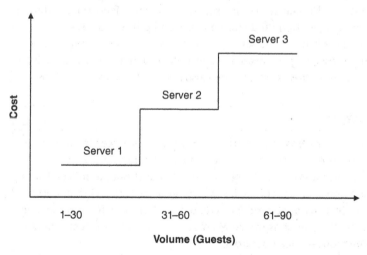

Figure 6.7 Sales Volume and Step Costs

Direct and Indirect (Overhead) Costs

One reason foodservice operators carefully consider costs is to assign them to specific operating areas within their businesses. For example, if the operation serves alcohol, the costs of providing mixers, fruit juices, and vegetables used for drink making is typically designated as a bar cost, even if the items were purchased as part of food costs.

When a cost can be directly attributed to a specific area or profit center within a business, it is known as a **direct cost**. Direct costs usually (but not always) increase with higher sales volumes. The most common direct costs in a restaurant relate to the food and beverages being sold. On a restaurant's income statement, these expenses are usually identified as "Cost of sales" (see Chapter 3).

Key Term

Direct cost: A cost that most often increases as activity or the volume within a revenue center increases.

An **indirect cost** is one not easily assigned to a specific operating unit or department. Consider the accountant's position in a foodservice business that operates in three locations. That person is likely responsible for the cash control and financial systems for all three units. The cost of employing the accountant is considered an indirect cost because it is not easily attributable to any one unit.

Key Term

Indirect cost: A cost that cannot be readily assigned to a specific revenue center in a foodservice operation. Also referred to as an overhead cost.

Typically, indirect costs include those for administrative and general expenses, information systems, human resources, security, franchise fees, transportation, marketing, property operations and maintenance, and utility expenses.

Other indirect costs can include non-operating expenses such as rent and other facility occupation costs, property taxes, insurance, depreciation, amortization, interest, and income taxes. Indirect costs are often also referred to as "overhead" costs.

Technology at Work

LinkedIn is a popular website that offers users several services in addition to advertising their site visitor's own profiles and employment histories.

One of LinkedIn's lesser-known features is its educational products. These include lessons on a variety of topics including accounting. One important feature of the instructional products created by LinkedIn for costing and accounting is that the lessons utilize Microsoft Excel, a spreadsheet program familiar to most foodservice operators.

To review some of LinkedIn's innovative lesson plans related to accounting and to indirect costs specifically, enter "LinkedIn activity-based costing in Excel" and then enter "LinkedIn assigning indirect costs" in the search bar and view the results.

Controllable and Non-Controllable Costs

Some costs can be controllable, and other costs can be non-controllable. **Controllable costs** are those costs over which a property-level manager has primary control. In contrast, non-controllable costs (see Chapter 3) are those costs that a property-level manager cannot typically control in the short term.

Key Term

Controllable costs: Those costs over which on-site foodservice managers have primary control.

In most foodservice operations, managers are only held responsible for unit profits remaining after subtracting the expenses they can directly control. In contrast, on-site managers cannot directly affect most non-controllable costs. In a restaurant, examples of controllable costs are costs of sales and labor. This is also a reason why income statements created using a USAR format (see Chapter 3) clearly indicate controllable and non-controllable expenses.

All costs can at least be partially controlled by the business owners because they can elect to continue, close, or expand their operations. On-site managers and their supervisors make a distinction, however, between controllable and non-controllable costs to indicate items for which managers can be held directly responsible and other items for which the business's owners are most responsible.

To illustrate the importance of understanding controllable and non-controllable costs, consider Stewart, who operates a neighborhood tavern/sandwich shop. Most of his revenue comes from the sale of beer, sandwiches, and his specialty wood-fired pizzas. Stewart can decide, on a weekly or monthly basis, the amount he will spend on advertising. Advertising expense, then, is under Stewart's direct control and is considered a controllable cost.

Some of Stewart's expenses, however, are not under his control. Licensing fees are a familiar form of non-controllable cost. The state in which Stewart operates charges an annual licensing fee for all operations that serve alcoholic beverages. If the state in which he operates increases this licensing fee, Stewart must pay the higher amount. In this situation, the alcoholic beverage licensing fee is a non-controllable cost because it is beyond Stewart's immediate control.

As a second example, assume a quick-service property sells takeout chicken, and the unit is part of a nationwide franchised chain of such stores. Each month, the store is charged a $600 advertising and promotion fee by the chain's regional headquarters' office. The $600 is used to purchase television advertising time for the franchised brand. This $600 charge, set by the franchisor, is a non-controllable operating cost, even for the property's highest-level managers.

In each of these fixed cost examples, the unit operator will find that even the best control systems will not affect (reduce) some specific costs. As a result, experienced operators focus most of their attention on managing controllable, rather than non-controllable, costs.

Find Out More

Some hospitality costs are not as controllable as they first appear. If a foodservice operation plays background music, hosts live bands or Karaoke nights, or even allows DJs to set up in the dining or banquet rooms, the operation will be required to pay artists' royalties for the music it plays. Although the decision to play music is controllable, the royalty costs associated with playing that music are non-controllable.

Three groups represent those who produce music:

The American Society of Authors, Composers, and Publishers (ASCAP) is a membership association of over 200,000 U.S. composers, songwriters, lyricists, and music publishers of every kind of music. ASCAP was created by and is controlled by composers, songwriters, and music publishers, with a Board of Directors elected from its membership.

BMI is an American performing rights organization that represents more than 300,000 songwriters, composers, and music publishers in all genres of music. The non-profit-making company, founded in 1939, collects license fees on behalf of those American artists it represents.

SESAC is a performing rights organization with headquarters in Nashville, Tennessee. SESAC's repertory, once limited to European and gospel music, has diversified to include today's most popular music, including R&B, hip-hop, dance, rock classics, country, Latino, Contemporary Christian, jazz, and the television and film music of Hollywood.

To learn more about these organizations and how they collect fees for their member's work, enter the organization's name in your favorite browser and view the results.

Other Types of Costs

Other types of costs important to foodservice operators include:

✓ Joint costs
✓ Incremental costs
✓ Standard costs
✓ Sunk costs
✓ Opportunity costs

Joint Costs

Joint costs are closely related to overhead and cost allocation issues because they involve allocating costs to two (or more) departments or profit centers.

Key Term

Joint cost: A cost that benefits more than one product or profit center. Also referred to as a "shared" cost.

For example, in a large operation, an Executive Chef is typically responsible for the food preparation activities of the restaurant's dining room(s) and the banquet department even though these two areas are often considered separate profit centers. In this case, the Executive Chef's salary could be considered a joint or shared cost associated with each revenue-producing area.

Most direct costs are not joint costs, while many indirect costs are considered joint costs. As with overhead costs, one difficulty associated with the assigning of joint costs is determining a logical basis for the cost allocation. Commonly used joint cost allocation factors include guests served, hours worked, wages paid (for staffing-related joint costs), sales revenue generated, and hours of operation.

Incremental Costs

For many foodservice operators, the computation of incremental costs is very important. Incremental is a variation of the word, "increment," which means "to increase." **Incremental costs** can best be understood as the increased cost of "each additional unit," or even more simply, as the cost of "one more."

Foodservice operations are concerned about incremental costs. To illustrate, consider a pizza operation that offers "free" delivery service to customers in a nearby apartment complex. The store's managers know that the total cost of providing the pizza to customers must include the cost of the pizza itself and the expense incurred delivering the pizza. Both of these costs can be computed rather easily.

> **Key Term**
>
> **Incremental cost:** The total cost incurred due to an additional unit of product being produced or a guest being served.

If, however, the driver can be scheduled so that two pizzas (not one) are delivered in a single trip to the complex, the cost of delivery will not double. In fact, the incremental cost of delivering the second pizza during the same delivery trip will be small because it would include only the driver's additional delivery time and not the cost of transportation to and from the complex.

Pizza delivery is an excellent illustration of why managerial accountants concern themselves with the computation and management of incremental costs. They do so because, in many situations, the fixed costs required to deliver a service or product to "one more" can be relatively small, and profits made on the sale to "one more" can be relatively large.

Standard Costs

Managerial accountants and knowledgeable foodservice operators read their income statements to find out what their costs *are*. The best operators, however, also want to know what their costs *should be*. Remember that management's primary responsibility is not to eliminate costs; instead, it is to incur costs appropriate for the quality of products and services delivered to guests. When they do so, the business will prosper. If management focuses on reducing costs more than servicing guests properly, problems will certainly surface.

Guests cause businesses to incur costs. Operators must resist the temptation to think that "low" costs are good and "high" costs are bad. A foodservice operation generating $2,000,000 in revenue per year will undoubtedly have higher costs than the same size operation generating $1,000,000 in revenue per year. The reason is clear. The food products, labor, and equipment needed to sell $2,000,000 worth of food are much greater than that required to produce a smaller amount of revenue. If there are fewer guests, there are likely to be fewer costs, but fewer profits as well. Operators seek to understand their **standard costs**, simply defined as the costs that *should* be incurred given a specific level of volume.

Standard costs can be established for nearly all business expenses from insurance premiums to plate garnishes. For those with foodservice experience, understanding standard costs is easy because they already understand standardized recipes. Just as a **standardized recipe** describes how a menu item should be cooked and served, a standardized cost describes how much it should cost to prepare and serve the item. If the variation from the standard cost is significant, it should be of concern to management.

To illustrate standard costs, assume that a foodservice operator determines $10.00 is the standard cost of preparing and serving a 16-ounce rib-eye steak. The steak sells for $35.00. If, at the end of the month, the manager finds the actual cost of preparing and serving each steak is $15.00, rather than $10.00, profits will suffer. Clearly, the operator must investigate the reasons for the difference between the item's standard cost and its actual cost.

In a similar manner, however, if the operator found, at month's end, that the actual cost of providing the steak was $7.00, long-term profits are also likely to suffer. The reason: At $7.00 cost per steak sold, it is unlikely the operation is giving its guests the perceived "rib-eye value" intended by management. When costs are significantly *below* the standard established for them, reduced quality of products served or poor service levels are often the cause. Experienced operators know that significant variations either above or below standard costs should be carefully investigated.

> **Key Term**
>
> **Standard cost:** The estimated expense that occurs during the production of a product or performance of a service. Also referred to as an "expected" cost, a "budgeted" cost, or a "forecasted" cost.

> **Key Term**
>
> **Standardized recipe:** The instructions needed to consistently prepare a specified quantity of food or drink at an expected quality level.

An effective foodservice operator knows the most important question to be considered is *not* whether costs are high or low. The critical question is whether costs incurred are at or very near the standards established for them.

Managers can eliminate nearly all costs by closing an operation's doors. Obviously, however, when the door to expense is closed, the door to profits closes as well. Expenses, then, must be known in advance (a standard must be established), costs will be incurred, and these costs must then be monitored to allow the operator to correct significant variations from pre-identified standard costs.

Sunk Costs

A **sunk cost** is one that has already been incurred, and whose amount cannot be altered.

Information about a sunk cost should be disregarded when considering a future decision, and they are most often identified and considered when making decisions about asset replacement or acquisition.

Key Term

Sunk cost: Money that has already been spent and cannot be recovered.

To illustrate, assume a manager is responsible for real estate acquisition in a large chain of restaurants. The manager is considering the purchase of two different pieces of land on which to build a new unit. Neither piece of land is located near the manager's office, so the manager must take airline flights to examine the two sites.

The cost of visiting the closest site is $1,000 and the cost of visiting the second site is $2,000. After the manager has concluded the trips, both the $1,000 and the $2,000 must be considered sunk costs. That is, they are costs about which the manager can now do nothing. They were necessary expenses in the decision-making process about which piece of land is best. These costs should not now, however, be among the factors the manager should consider when determining which of the two sites would make the best location for the new restaurant.

Opportunity Costs

An opportunity cost is the opposite of an expense because it is the cost of *not* electing to take a course of action. Specifically, an **opportunity cost** is the cost of foregoing the next best alternative when making a decision.

For example, assume a foodservice operator has two choices, A and B, both having potential benefits or returns. If the operator selects A, he or she will lose the potential benefits from choosing B (the opportunity cost).

Key Term

Opportunity cost: The potential benefits that an investor misses out on when selecting one alternative course of action over another.

To illustrate, consider the operator who has some extra cash on hand for the month. The operator could decide to use the money to buy food inventory that will sit in the storeroom until it is sold, or he or she could decide to invest the money. If the operator elects to buy the excess inventory, the opportunity cost is the amount of money that would have been made if the cash were invested in an interest-bearing bank account.

What Would You Do? 6.1

"I can buy it at a great price," said Dan Flood who was talking about the Waterford Restaurant. The property was for sale, and Dan was meeting with Loralei Nystrom, his friend and an experienced restaurant manager. "It's losing about 5 cents on each dollar of sales now," continued Dan, "but I know we can turn that around."

Loralei considered Dan's proposal that they form a partnership, acquire the restaurant, and share the profits they planned to make. She knew that, before it was possible to share profits, they would actually have to make a profit. That meant, to go from losing 5 cents per dollar to making money, they would have to increase sales, reduce costs, or both. She mentioned that to Dan.

"Well," he replied, "I'm not sure we need to increase the sales at all. If we buy it at the right price, I think we just need to reduce our operating costs, and you can do that."

Assume you were Loralei. Why would it be important for you to know about this operation's fixed and non-controllable costs under the arrangement proposed by Dan? Why would it be important to know about the operation's likely variable and controllable costs if you operated it?

Break-even Analysis

Experienced managerial accountants know that some accounting periods are more profitable than others in businesses that experience "busy" and "slow" periods. For example, a ski resort may experience tremendous food sales during the ski season but then have a greatly reduced volume (or may even close) during the summer. Similarly, a country club manager in the Midwestern United States knows revenue from greens fees, golf outings, and in-house food and beverage sales are high in the summer months. However, they also know the golf course will likely be closed for several winter months.

Experienced foodservice operators who understand costs know that costs as a percentage of sales are generally reduced when sales are high, and these percentages increase when sales volume is lower. The result, in most cases, is greater profits during high-volume periods and lesser profits in lower-volume periods. This relationship among volume, costs, and profits is easier to understand when examined graphically as shown in Figure 6.8.

The *x* axis in Figure 6.8 represents sales volume: the number of covers (guests) served or dollar volume of sales. The *y* axis represents the costs associated with generating these sales. The Total Revenues line starts at 0 because if no guests are served, no revenues are generated. The Total Costs line starts further up the *y* axis because fixed costs are incurred even if no sales are made.

The point at which the two lines cross is called the **break-even point** in which operating expenses exactly equal revenue. Stated another way, when sales volume equals the sum of its total fixed and variable costs, its break-even point has been reached. Below the break-even point, costs are higher than revenues and losses occur. Above the break-even point, revenues exceed the

Key Term

Break-even point: The sales point at which total cost and total revenue are equal. There is no loss or gain for a business.

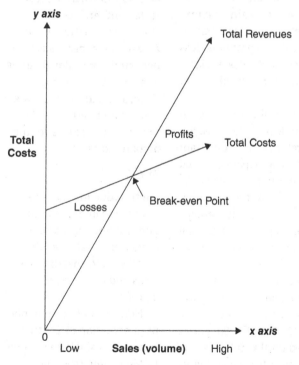

Figure 6.8   Cost/Volume/Profit (Break-even)

sum of the fixed and variable costs required to make the sales, so profits are generated.

Computation of Break-even Point

Most foodservice operators would like to know the break-even point of their businesses on a daily, weekly, or monthly basis. In effect, by determining the break-even point, the operator is answering the question, "How much sales volume must be generated before I begin to make a profit?" Beyond the break-even point, the operator wants answers to another question: "How many sales dollars and volume must I generate to make my *target* profit level?"

To answer these questions, operators must calculate their break-even points. Doing so allows them to predict the sales dollars and volume required to achieve a break-even point or desired profit based on known costs.

Before a break-even point analysis can be calculated, a **contribution margin income statement** must be developed. A contribution margin income statement shows profit and loss (P&L) items relative to terms of sales, variable costs, **contribution margin**, fixed costs, and profit.

To illustrate, consider again the Golden Dragon Restaurant. In the Mixed Costs section of this chapter, the operator of this restaurant used the high/low method to separate mixed costs into their variable and fixed components. As was illustrated in Figure 6.4, the operation served 10,000 guests in October, and its variable cost per guest was $5.00. This resulted in total variable costs of $50,000 ($5.00 × 10,000 guests = $50,000).

In addition, the operation's fixed costs were $60,000. If it is assumed that sales per guest (guest check average) was $12.50, then total sales were $125,000 ($12.50 × 10,000 guests served = $125,000). Based on this information, the Golden Dragon's operator has converted its P&L statement to a contribution margin income statement as shown in Figure 6.9.

Key Term

Contribution margin income statement: An income statement variation in which all variable expenses are deducted from sales to arrive at a contribution margin. Then, all fixed expenses are subtracted to arrive at the net profit or net loss for the accounting period.

Key Term

Contribution margin: The dollar amount remaining after subtracting variable costs from total sales that *contributes* to covering fixed costs and providing for a profit.

Note: Another commonly used definition of CM that relates to a single menu item is "Menu item revenue – Menu item product costs = CM."

Total Sales	$125,000	Sales Per Guest	$ 12.50
Variable Costs	50,000	Guests Served	10,000
Contribution Margin	75,000		
Fixed Costs	60,000		
Before Tax Profit	15,000		
Taxes (40 %)	6,000		
After-tax Profits	$ 9,000		

Figure 6.9   Golden Dragon Contribution Margin Income Statement for October

As shown in Figure 6.9, the contribution margin calculation for the Golden Dragon is:

Total Sales – Variable Costs = Contribution Margin

Or

$125,000 – $50,000 = $75,000$

Using the same data, the operator can also view the contribution margin income statement in terms of per guest and percentage sales, variable costs, and contribution margin (CM) as shown in Figure 6.10.

Notice the boxed information in Figure 6.10 includes per guest and percent calculations. These include selling price (SP), variable costs (VC), and contribution margin (CM). Note also that fixed costs are not calculated as per unit or as a percentage of sales. The reason: Fixed costs do not vary as sales volume increases or decreases.

		Per Guest		**Percent**
Total Sales	$125,000	SP	$12.50	100
Variable Costs	50,000	VC	5.00	40
Contribution Margin	75,000	CM	7.50	60
Fixed Costs	60,000			
Before-tax Profit	15,000	Guests served = 10,000		
Taxes (40%)	6,000			
After-tax Profits	9,000			

Figure 6.10   Golden Dragon Contribution Margin Income Statement with Per Guest and Percent Calculations

To calculate these numbers, operators use the following steps:

Step 1: Divide total sales, variable costs, and contribution margin by the number of guests to determine per guest values. Then, calculate CM/guest.

SP/guest = $125,000 / 10,000 guests = $12.50
VC/guest = $ 50,000/10,000   guests = $ 5.00
CM/guest = $ 75,000/10,000   guests = $ 7.50

SP/guest − VC/guest = CM/guest

Or

$12.50 − $5.00 = $7.50

Step 2: Divide VC/guest by SP/guest, and CM/guest by SP/guest to get percentage values. Then, calculate CM%.

SP% = 100%
VC% = $5.00/$12.50 = 40%
CM% = $7.50/$12.50 = 60%

SP% − VC% = CM%

Or

100% − 40% = 60%

Once the Golden Dragon P&L statement has been converted to a contribution margin income statement and per guest values and percentages are calculated, its operator can determine the operational break-even point and sales required to achieve desired profit. This can be done based on dollar sales and the number of guests required to do so.

To determine the dollar sales required to break even, the operator uses the following formula:

$$\frac{\text{Fixed Costs}}{\text{Contribution Margin \%}} = \text{Break-even Point in Sales Dollars}$$

Or

$$\frac{\$60,000}{0.60} = \$100,000$$

As seen above, the Golden Dragon must generate $100,000 in sales per month *before* it begins to make a profit. At a sales volume of less than $100,000, there is an operating loss.

To determine the number of guests to be served at the break-even point, the operator uses the following formula:

$$\frac{\text{Fixed Costs}}{\text{Contribution Margin per Guest}} = \text{Break-even Point in Guests Served}$$

Or

$$\frac{\$60,000}{\$7.50} = 8,000 \text{ Guests}$$

Now, assume that the Golden Dragon's operator decided that in July they will plan for \$12,000 in after-tax profits. To determine sales dollars and covers needed to achieve this goal, the operator uses the following formula:

$$\frac{\text{Fixed Costs} + \text{Before-Tax Profit}}{\text{Contribution Margin \%}} = \text{Sales Dollars to Achieve Desired After-Tax Profit}$$

This operator knows that the after-tax-profit goal is \$12,000, but the preceding formula calls for *before-tax profit*. To convert after-tax profit to before-tax profit, the operator must compute the following:

$$\frac{\text{After-Tax Profit}}{1.00 - \text{Tax Rate}} = \text{Before-Tax Profit}$$

Or

$$\frac{\$12,000}{1.00 - 0.40} = \$20,000$$

Since the operator knows that the before-tax profit goal is \$20,000, he or she can calculate the sales dollars needed to achieve the desired after-tax profit as follows:

$$\frac{\text{Fixed Costs} + \text{Before-Tax Profit}}{\text{Contribution Margin\%}} = \text{Sales Dollars to Achieve Desired After-Tax Profit}$$

Or

$$\frac{\$60,000 + \$20,000}{0.60} = \$133,333.33$$

As seen above, the operation must generate \$133,333.33 in sales in July to achieve its desired after-tax profit of \$12,000.

To calculate the number of guests that must be served to generate the profit, the Golden Dragon operator uses the following formula:

$$\frac{\text{Fixed Costs} + \text{Before-Tax Profit}}{\text{Contribution Margin per Guest}} = \text{Number of Guests to Achieve Desired After-Tax Profit}$$

Or

$$\frac{\$60,000 + \$20,000}{\$7.50} = 10,667 \text{ Guests (rounded up)}$$

Note that the number of guests was rounded up from 10,666.67 to 10,667.

After operators fully understand these break-even concepts, they can use them to predict sales levels for break-even or after-tax profits based on selling prices, fixed costs, variable costs, and contribution margin. They can then make changes in selling prices and costs to improve their ability to break-even and achieve desired profit levels.

Minimum Sales Point

All operators should know how to compute their break-even points, and many also can compute a **minimum sales point (MSP)**: the dollar sales volume required to justify staying open for a given period of time.

For foodservice operators, the information necessary to compute an MSP is:

1) Food cost percentage
2) Minimum payroll cost for the time period
3) Variable cost percentage

> **Key Term**
>
> **Minimum sales point (MSP):** The dollar amount of sales needed for a business to profitably remain open during a defined time period.

Fixed costs are eliminated from the calculation because, even if the volume of sales equals zero, the fixed costs still exist and must be paid.

To illustrate the use of a minimum sales point, consider Saanvi, the owner of an operation featuring Bengali Indian cuisine. Saanvi wants to determine whether she should close the operation at 10:00 p.m. or 11:00 p.m. based on the sales volume necessary to justify staying open the additional hour. She can make this calculation because she knows her food cost equals 40%, her minimum labor cost to stay open for the extra hour equals $150, and her other variable costs (taken from her income statement) equal 30%.

In calculating MSP, her food cost % + variable cost % is called her **minimum operating cost**.

Saanvi can apply the MSP formula as follows:

> **Key Term**
>
> **Minimum operating cost:** The least dollar amount of cost of sales plus variable costs incurred by a foodservice operation during a defined time period.

$$\frac{\text{Minimum Labor Cost}}{1 - \text{Minimum Operating Cost}} = \text{MSP}$$

Or

$$\frac{\text{Minimum Labor Cost}}{1.00 - (\text{Food Cost}\% + \text{Variable Cost}\%)} = \text{MSP}$$

Or

$$\frac{\$150}{1.00 - (0.40 + 0.30)} = \$500$$

If Saanvi can achieve a sales volume of $500 in the 10:00 p.m. to 11:00 p.m. time period, she should stay open. If this level of sales is not feasible, she should consider closing the operation at 10:00 p.m.

Of course, some operators may not have the authority to close the operation, even when remaining open is not particularly profitable. Corporate policy, contractual hours, promotion of a new unit, competition, and other factors must be considered before the decision is made to modify operational hours regardless of projected volume levels.

Find Out More

One of a foodservice operator's most important decisions relates to their operation's opening and closing days and times. Costs, of course, are of major concern when making this decision, but so are other important factors.

In some cases, an operation's menu helps dictate the hours of operation. For example, an operation featuring breakfast items must be open in the mornings. In other cases, state or local laws may apply. This is particularly true of those operations that serve alcoholic beverages.

In some locations, staffing may be an issue. A foodservice operation must be able to properly staff its business during all operating hours.

To review some of these key issues, as well as others that impact the decision of when to open and close a foodservice operation, enter "how to establish restaurant operating hours" in your favorite search engine and review the results.

Technology at Work

Break-even analysis is the process of determining an operation's break-even point. It requires considering fixed costs, variable costs, price per guest, and the number of guests.

Break-even analysis is helpful when:

1) You want to identify the number of guests needed to reach the break-even point. For example, when you're launching a new set of menu items, and

(Continued)

> you want to know the minimum number of guests to be served to recover your fixed and variable costs.
> 2) You know the number of guests you will likely serve but want to decide on the right selling price (guest check average) for your menu items. For example, you have established the target number of guests, but are finding it a challenge to establish the right menu prices for your items.
>
> Some companies offer free-to-access break-even software that helps in these two areas. To examine some of the software options available to foodservice operators who want assistance in determining their break-even points, enter "break-even calculator" in your favorite search engine and review the results.

Foodservice operators who understand the various costs they will incur in managing their businesses are in an excellent position to control and manage these costs. The control of costs is important because an operation's revenue minus its cost yields its profits (or loss!).

Management of an operation's revenue is equally important. Foodservice operators directly impact their revenue with the pricing strategies and tactics they use. Since pricing for profits is so important in a foodservice operation, the proper pricing of foodservice menu items is the sole topic of the next chapter.

What Would You Do? 6.2

"I think we have to do it," said Blanca, the Director of Operations for the five-unit Hungry Hugo's pizza chain. Blanca was talking to Hugo Acosta, the chain's founder.

Blanca and Hugo were discussing that their competitors had extended their hours of operation. Instead of closing at 10 p.m. or 11 p.m. as they had in the past, increasing numbers of competitors were extending their hours to midnight or even 2 a.m.

Since its founding 20 years earlier, the Hungry Hugo's chain units closed at 11 p.m., even on weekends.

"I don't know," replied Hugo. "Our sales between 10 p.m. and 11 p.m. are a good bit lower already than they are between 7 p.m. and 9 p.m. I'm not sure staying open until midnight will generate much more business. How much will it cost us to do it!"

"Well," replied Blanca, "I think if we don't do it, we are in danger of being left behind by our competitors. Once customers are used to going elsewhere it can be hard to get them back, even though our product quality is much better than the others!"

Assume you were Hugo. How important would it be for you to understand the break-even point in your operation? How important would it be to understand your minimum sales point? Explain your answer.

Key Terms

Activity-based costing
Mixed cost
Step costs
Direct cost
Indirect cost
Controllable costs
Joint cost

Incremental cost
Standard cost
Standardized recipe
Sunk cost
Opportunity cost
Break-even point

Contribution margin
  income statement
Contribution margin
Minimum sales
  point (MSP)
Minimum operating cost

Operator's 10-Point Tactics for Success Checklist

Evaluate your need for, and the current status of, each of the following operational tactics. For those tactics you think are important, but not yet in place, develop an action plan for its implementation including who will be responsible for the tactic's completion and the target date by which it should be completed.

Tactic	Don't Agree (Not Done)	Agree (Done)	Agree (Not Done)	Who Is Responsible?	Target Completion Date
				If Not Done	
1) Operator understands that the term "cost" actually refers to a large number of different types of operating expenses.	____	____	____		
2) Operator recognizes the difference between a variable expense and a fixed expense.	____	____	____		
3) Operator can identify those expenses that are classified as "mixed."	____	____	____		
4) Operator recognizes the unique characteristics of a "step" cost.	____	____	____		
5) Operator can distinguish between a direct and indirect (overhead) cost.	____	____	____		

(Continued)

				If Not Done	
Tactic	Don't Agree (Not Done)	Agree (Done)	Agree (Not Done)	Who Is Responsible?	Target Completion Date
6) Operator can distinguish between a controllable and a non-controllable cost.	___	___	___		
7) Operator understands the differences between joint, incremental, standard, sunk, and opportunity costs.	___	___	___		
8) Operator can prepare a contribution margin income statement.	___	___	___		
9) Operator can prepare a break-even analysis when operating costs are known.	___	___	___		
10) Operator can calculate a minimum sales point (MSP) when minimum operating costs are known.	___	___	___		

7

Profitable Pricing

What You Will Learn:

1) The Relationship Between Prices and Profits
2) The Factors Affecting Foodservice Prices
3) The Methods Used to Price Menu Items
4) How to Evaluate Pricing Efforts in Foodservice Operations

Operator's Brief

In this chapter, you will learn that properly pricing menu items is an important skill for all foodservice operators even though prices are viewed differently by sellers and buyers. For operators, costs and profits are critical when establishing menu prices, but guests are concerned about the value they receive. Therefore, in addition to pricing menu items to be profitable, foodservice operators must also ensure menu prices communicate real value to their guests.

Numerous factors influence prices foodservice operators charge for their menu items. Among the most important are economic conditions, competition, level of service, type of guest, and product quality and costs. Additional pricing factors include portion size, delivery style, meal period, location, and bundling: a pricing strategy that combines multiple menu items that are then sold at a price lower than that of the bundled items purchased separately.

When establishing menu prices, some operators use a product cost-based approach. They believe a menu item's production cost relative to its price is of most concern. When using this pricing approach, menu items with lower product cost ratios are thought to be more desirable to sell than those with higher product cost ratios. Other operators use a more profit-oriented approach to establish their menu prices. Menu items providing high profits-per-sale are considered more desirable to sell than those with lower profit levels.

(Continued)

Regardless of the pricing approach used, foodservice operators should regularly evaluate their menus to identify items that are the most popular and profitable. Then they can modify or even eliminate poor selling or unprofitable items and better promote those that are the most popular and most profitable.

CHAPTER OUTLINE

Pricing for Profits
    The Importance of Price
    The Operator's View of Price
    The Guest's View of Price
Factors Affecting Menu Pricing
    Economic Conditions
    Local Competition
    Level of Service
    Type of Guest
    Product Quality
    Portion Size
    Delivery Method
    Meal Period
    Location
    Bundling
Methods of Food and Beverage Pricing
    Cost-based Pricing
    Contribution Margin-based Pricing
Evaluation of Pricing Efforts
    Menu Engineering
    Menu Modifications

Pricing for Profits

Foodservice operators must price their menu items to help ensure the long-term profitability of their businesses. Experienced foodservice operators know that, if menu prices are too low, an operation may be popular but not profitable. Alternatively, if menu prices are too high, the popularity of a foodservice operation will likely suffer because few guests regularly use the foodservice operation. As a result, operators who establish menu item prices must understand how to do so effectively.

The Importance of Price

Price plays a large marketing role in interactions between a foodservice operation and its

Key Term

Price (Noun): A measure of the value given up (exchanged) by a buyer and a seller in a business transaction.

For example: "The price of the chicken sandwich combo meal is $9.95."

guests. To best understand the importance of price, foodservice operators must understand that the term as used in the foodservice industry has two separate definitions.

Note that in both uses (noun or verb) the concept of an *exchange* between a buyer and a seller is important. The foodservice operator gives up (exchanges) a menu item as it is purchased, and the foodservice guest gives up (exchanges) the item's selling price for the menu item.

Price (Verb): To establish the value to be given up (exchanged) by a buyer and a seller in a business transaction.

For example: "We need to price the chicken sandwich combo meal."

Foodservice operators can typically charge any price they want to charge, but potential guests can accept or reject the operator's opinion that prices charged are fair and provide good value to consumers. In fact, a foodservice operation itself can be selected by a foodservice guest based primarily on the prices charged and the guest's perceptions of those prices.

To best understand pricing in the foodservice industry, it is important to first recognize that price is viewed very differently from the perspectives of an operator (seller) and a guest (buyer).

The Operator's View of Price

In most cases, operators decide what products they sell, where they will sell them, and how they communicate their prices to potential guests. Foodservice operators can propose their prices, but they also must face the possibility that potential guests may *not* support their value propositions.

In the foodservice industry, price and value are related concepts. It is often stated that the value of any item is equal to what a buyer will pay for it. If this is true, when a sale is *not* made, the buyer either believed the item was not worth the asking price or a lower cost alternative was available that was also considered worth its asking price.

From the perspective of an operator, a fair price should be an amount equal to the operator's incurred costs plus a reasonable desired profit.

Stated mathematically, that concept is:

Item Cost + Desired Profit = Selling Price

When calculating item costs, operators consider the estimated costs of food, beverage, labor, and all other costs required to operate their businesses. Hospitality business owners and accountants can generally calculate these costs quite accurately.

The question of "*What is a reasonable desired profit?*" is more subjective. In the foodservice industry, an operation's **profit margin** is the

Key Term

Profit margin: The amount by which revenue in a foodservice operation exceeds its operating and other costs.

percentage of each buyer's revenue dollar the operation retains as profit. The higher the profit margin, the more profitable is the operation.

In most cases, a foodservice operator wants to generate a reasonable profit margin because doing so is critical to staying in business and receiving a fair return for the investment risks in the business. Therefore, consideration of operating costs incurred and profit desired are the two most important concerns as operators establish menu prices.

The Guest's View of Price

Guests desire a good **value** for the products and services they purchase. If they do not feel they have received good value, the guests are unlikely to be satisfied and will not return to the business. Foodservice operators must remember that value is determined by the buyer, not the seller, in any business transaction.

Key Term

Value: The amount paid for a product or service compared to the buyer's view of what they receive in return.

For example, a foodservice operator may sincerely believe that an item's quality, quantity, and delivery method will provide good value to guests if it is sold for $29.95. If, however, too few customers share that view of value, the menu item will not be frequently sold. Stated mathematically, a buyer's perception of value is expressed as:

$$\text{Buyer s Perceived Benefit(s)} - \text{Price} = \text{Value}$$

When making a purchase, buyers want to receive *more* value than the value of what they are giving up. Stated another way, there are three possible buyer reactions to any seller's proposed selling price as shown in Figure 7.1.

When buyers believe the benefit they will receive is less than zero (less than what they give up in exchange), they generally will not buy. When the benefit is greater than zero, guests are very likely to buy. When the perceived benefit is equal to "0," buyers are often indifferent to the purchase. In this scenario, if no other alternatives are available, they may make a purchase decision. If alternatives are available, buyers will likely consider these alternatives before making purchase decisions.

Buyer's Assessment	Purchase Decision
1. Perceived Benefit − Price = A value less than "0"	Do not buy
2. Perceived Benefit − Price = A value equal to "0"	Do not buy in most cases
3. Perceived Benefit − Price = A value greater than "0"	Buy

Figure 7.1 Buyer's Assessment of a Seller's Value Proposition

To best understand how buyers' perceived benefit assessments directly impact their purchasing
decisions, operators must first understand the concept of **consumer rationality**.

Consumer rationality assumes that buyers
consistently exhibit reasonable and purposeful
behavior. That is, buyers generally make purchase decisions based solely on their belief that it
benefits them to do so.

For foodservice operators, the acceptance of the concept of consumer rationality involves a willingness to look beyond the obvious and attempt to understand
exactly how buyers believe they will benefit from a business transaction.

In some cases, this is not easy. First, foodservice operators must resist the temptation to declare their guests are irrational (e.g., when operators criticize guests
who state the operator's prices are "too high!").

All buyers like low prices, but what they seek most is value. Most foodservice
guests are indifferent to the actual costs of operating a foodservice business and,
therefore, they are indifferent to an operator's profit margin. Also, rational buyers *do
not* automatically equate a seller's price with the amount of value they (buyers)
receive in an exchange. In fact, conventional wisdom advises them not to do so.
From a common sense and even from a legal perspective, buyers assessing a seller's
value proposition are cautioned not to trust sellers. As a result, *caveat emptor,* the
Latin phrase for "let the buyer beware," is known and understood by most consumers.

Since many buyers may be skeptical about a seller's initial value proposition, the
foodservice operator's responsibility when pricing menu items is to ensure guests
understand the answers to questions like *"What do I get?"* and *"Why is it of value?"*
just as much as they understand the actual prices they will pay. Only then can
buyers, who are increasingly sophisticated and web-savvy consumers, learn they
will consistently receive *more* than the worth of the money they must pay when
making purchase decisions.

A foodservice operator's primary motivation to recover operating costs and generate a profit is very different from their guests' motivation to optimize the value
received for the prices they pay. Therefore, while operators are concerned about
their costs and profits, a marketing view is that they be *more* concerned about
utilizing price as a method of communicating excellent value to their guests.

Factors Affecting Menu Pricing

When foodservice operators find that profits are too low, they frequently question
whether their prices (and therefore their revenues) are too low. Remember that
the terms "revenue" and "price" are not the same thing. "Revenue" is the amount

of money spent by all guests, and "price" refers to the amount charged for one menu item. Total revenue for that menu item is generated by the following formula:

$$Price \times Number\ Sold = Total\ Revenue$$

There are two components of total revenue. Price is one component, and the other is the number of items sold. Note: Generally (but not always), as an item's selling price increases, the number of that items sold will decrease.

Experienced foodservice managers know that increasing prices without giving added value to guests does result in higher prices but, frequently, lower total revenue because of a reduction in the number of guest purchases. For this reason, menu prices must be evaluated based on their impact on the number of items sold as well as their actual selling price.

While paying attention to costs and profits is always important, other factors directly affect the prices operators charge for the menu items they sell. Among the most important of these are:

✓ Economic conditions
✓ Local competition
✓ Level of service
✓ Type of guest
✓ Product quality
✓ Portion size
✓ Delivery style
✓ Meal period
✓ Location
✓ Bundling

Economic Conditions

The economic conditions existing in a local area or even the entire country can significantly impact prices operators can charge for menu items.

A local robust and growing economy generally enables foodservice operators to charge higher prices for items they sell. In contrast, a local economy in recession and/or weakened by other events can limit an operator's ability to raise or maintain current prices in response to rising product costs.

In most cases, foodservice operators cannot directly influence the strength of their local economies. They can and should, however, monitor local economic conditions and carefully consider these conditions when establishing menu prices.

Local Competition

The prices charged by an operation's competitors can be important, but this factor is sometimes too closely monitored. It may seem to some operators that their average guest is only concerned with low price and nothing more. However, small price variations generally make little difference in the buying behavior of the average guest.

For example, if a group of young professionals goes out for pizza and beer after work, the major determinant is not likely whether the selling price for a small pizza is $16.95 in one operation or $18.95 in another. Other factors including quality and location become important.

The selling prices of potential competitors are of concern when establishing a selling price, but experienced operators understand that a specific operation can sell a product of lower quality for a lesser price. While competitors' prices can help an operator arrive at their properties' own selling prices, it should not be the only determining factor.

The most successful foodservice operators focus on building guest value in their own operations and not on attempting to mimic the efforts of competitors. Even though operators may believe customers only want low prices, remember that consumers often associate higher prices with higher quality products and, therefore, provide a better price/value relationship.

Level of Service

The service levels an operation provides directly affects the prices the operation can charge. Many guests expect to pay more for the same product when service levels are higher. For example, a can of soda sold from a vending machine is generally less expensive than a similar sized soda served by a service staff member in a sit-down restaurant.

Service levels can impact pricing directly and, as the personal level of service increases, selling prices may also increase. Personal service ranges from the delivery of products to one's home to the decision to quicken service by increasing the number of servers in a busy dining room (which improves service quality by reducing the number of guests each server assists).

These examples should not imply that extra income from increased menu prices is necessary to pay for extra labor required to increase service levels. Guests are willing to pay more for increased service levels. However, higher prices must cover extra labor costs and provide for extra profit as well. Many hospitality operators can survive and thrive over the years because of an uncompromising commitment to high levels of guest service, and they can charge menu prices reflecting enhanced service levels.

Type of Guest

All guests want good value for the money they spend. However, some guests are less price sensitive than others, and the definition of what represents good value can vary by the clientele served. Consider the pricing/purchasing decisions of convenience store customers across the United States. In these facilities, food products such as pre-made sandwiches, fruit, drinks, and cookies are often sold at relatively high prices. The customers in these stores most desire speed and convenience, and they are willing to pay premium prices for their purchases.

Similarly, guests at a fine-dining steakhouse restaurant are less likely to respond negatively to small variations in drink prices than are guests at a neighborhood tavern. A thorough understanding of exactly who the potential guests are and what they value most is critical to the ongoing success of foodservice operators as menu prices are established.

Product Quality

A guest's quality perception of a menu item can range from very low to very high, and perceptions are most often the direct result of how guests view an operation's menu offerings. These perceptions are directly affected by a menu item's quality, and they should never be shaped by the guest's view of an item's wholesomeness or safety. All foods must be wholesome and safe to eat. Guests' perceptions of quality are based on numerous factors of which only one is the quality of actual raw ingredients. Visual presentation, stated or implied ingredient quality, portion size, and service level are additional factors that impact a guest's view of overall product quality.

To illustrate, consider that, when most foodservice guests think of a "hamburger," they think of a range of products. A "hamburger" may include a rather small burger patty placed on a regular bun, wrapped in waxed paper, served in a sack, and delivered through a drive-thru window. If so, guests' expectations of this hamburger's selling price will likely be low.

If, however, the guests think about an 8-ounce "**Wagyu beef**-burger" with avocado slices and alfalfa sprouts on a fresh-baked, toasted, and whole-grain bun and served for lunch in a white-tablecloth restaurant, the purchase price expectations of the guests will be much higher.

A foodservice operator can select from a variety of quality levels and delivery methods when developing product specifications, and as they

Key Term

Wagyu beef: Beef from a Japanese breed of cattle that is highly prized for its marbling and flavor. In the Japanese language, "Wa" means Japanese, and "gyu" means cow.

plan their menus and establish their prices. The decisions they make will have a direct impact on menu pricing.

For example, if a bar operator selects an inexpensive Bourbon to make whiskey drinks, they will likely charge less for whiskey drinks made than another operator selecting a better (more expensive) brand. Guest perceptions of the value received from those buying the lower-cost whiskey drinks will likely be lesser than guests served a higher-quality product.

To be successful, foodservice operators should select the product quality levels that best represents the anticipated desires of their **target market,** and their operations' own pricing and profit goals.

Key Term

Target market: The group of people with one or more shared characteristics that an operation has identified as most likely customers for its products and services.

Portion Size

Portion size most often plays a large role in determining a menu item's price. Great chefs are fond of saying that people "Eat with their eyes first!" This relates to presenting food that is visually appealing, and it also impacts portion size and pricing.

A pasta entrée filling an 8-inch plate may be lost on an 11-inch plate. However, guests receiving the entrée on an 8-inch plate will likely perceive higher levels of value than those receiving the same entrée on an 11-inch plate even though the portion size and cost in both cases are identical.

Portion size is a function of both product quantity *and* presentation. Many successful cafeteria chains use smaller-than-average dishes to plate their food. For their guests, the image of price to value when dishes appear full comes across loud and clear.

In some foodservice operations, and particularly in those that are "all-you-care-to-eat" facilities, the previously mentioned principle again holds true. The proper dish size is just as critical as the proper-sized scoop or ladle when serving the food. Of course, in a traditional table service operation, an operator must carefully control portion sizes because the larger the portion size, the higher the product costs.

Many of today's health-conscious consumers prefer lighter food with more fruit and vegetable choices. The portion sizes of these items can often be boosted at a fairly low increase in cost. At the same time, average beverage sizes are increasing as are portion sizes of many side items such as French fries. If these items are lower-cost items, this can be good news for the operator. However, it is still important to consider the costs of larger portion sizes.

Every menu item to be priced should be analyzed to help determine if the quantity (portion size) being served is the "optimum" quantity. Operators would, of course, like to serve that amount, but no more. As will be addressed in detail in Chapter 8 (Food and Beverage Cost Control), the effect of portion size on menu prices is significant, and back-of-house staff should establish and maintain control over desired portion sizes.

Find Out More

Foodservice operators have a variety of choices when selecting software programs designed to help calculate the cost of pricing their various menu items. These menu management programs allow operators to insert their standardized recipes, portion sizes, and the cost of the ingredients used to make the recipes, and then the items' portion costs are automatically calculated.

Advances in menu management software continue to occur rapidly. Increasingly, foodservice operators look for programs providing options and then design their own pricing and sales tracking processes.

To stay current with newly developed menu pricing software and apps, enter "menu item pricing software" in your favorite search engine and review the results.

Delivery Method

Delivery method has become an increasingly important factor in establishing menu prices. There are essentially four ways foodservice operators deliver purchased menu items to their guests:

1) Dine-in service: Menu items are delivered to guests at their on-site tables or other seating areas.
2) Pick up/carryout: Guests receive their menu items from drive-through windows or pick up their items from designated on-site carry out areas.
3) Operator direct delivery: Menu items are delivered to guests by an operation's own delivery employees.
4) Third-party delivery: Menu items are delivered to guests by **third-party delivery** partners selected by the operation.

Key Term

Third-party delivery: The use of a smartphone or computer application that allows customers to browse restaurant menus, place orders, and have them delivered to the customer's location. In nearly all cases, the requested orders are delivered by independent contractors retained by the company operating the third-party delivery app.

The delivery style with the greatest impact on menu prices is third-party delivery. From the perspective of guests, the decision to utilize a third-party delivery company such as Grubhub,

DoorDash, or Uber Eats means the guest is placing important value on convenience as well as on the menu items they choose.

When guests order directly from a third-party delivery company, they actually pay for:

✓ The selling prices of their selected menu items
✓ A service fee charged by the delivery company for providing the service
✓ A delivery fee for food delivered
✓ A gratuity; an optional tip for the delivery driver

The COVID-19 era saw explosive growth in third-party delivery companies as many restaurants either closed indoor dining areas or severely restricted their inside seating capacities. However, the services of third-party delivery companies are not free to restaurants.

Depending upon the specific arrangement made between a foodservice operation and its third-party delivery partner(s), the operation will pay between 10% and 30% of a guest's total bill to the third-party delivery company. For example, if a foodservice operation has a customer who utilizes a third-party delivery app, and who purchases $100 worth of menu items from the operation, the operation will receive only $70.00 to $90.00 from the third-party delivery company.

Astute readers will recognize that the foodservice operation pays a significant fee to satisfy their customers' desire for convenience. Many foodservice operators charge different (higher) menu prices when items are ordered from a third-party delivery app or avoid the use of third-party delivery services entirely.

Meal Period

In some cases, guests pay more for an item served in the evening than for that same item served during a lunch period. Sometimes, this is the result of a smaller "luncheon" portion size, but in other cases the portion size and service levels may be the same in the evening as earlier in the day. This is true, for example, in buffet restaurants that charge a different price for lunch than for dinner. Perhaps operators expect those on lunch break to spend less time in the operation and will eat less. Alternatively, they may believe their guests simply spend less for lunch than dinner.

Foodservice operators must exercise caution in this area, however. Guests should clearly understand why a menu item's price changes with the time of day. If this cannot be answered to the guest's satisfaction, it may not be wise to implement a time-sensitive pricing structure.

Location

Some foodservice operators make their locations a major part of their key marketing messages, and location can be a major factor in price determination. This is illustrated, for example, by food facilities operated in themed amusement parks, movie theaters, and sports arenas.

Foodservice operators in these locations can charge premium prices because they have a monopoly on food sold to visitors. The only all-night diner on an interstate highway exit is in much the same situation. Contrast that with an operator who is just one of 10 similar seafood restaurants on a **restaurant row** in a seaside resort town. In this case, it is unlikely that an operation can charge prices significantly higher than its competitors based solely on location.

Key Term

Restaurant row: A street or region well-known for having multiple foodservice operations within close proximity.

One cannot discount the value of an excellent restaurant location, and location alone can influence price in some cases. Location does not, however, guarantee long-term success. Location can be an asset or a liability. If it is an asset, menu prices may reflect that fact. If location is a liability, menu prices may need to be lower to attract a sufficient clientele to ensure the operation achieves its total revenue and profit goals.

Bundling

Bundling refers to the practice of selecting specific menu items and pricing them as a group (bundle) so the single menu price of all the items purchased together is lower than if the items in the group were purchased individually.

Key Term

Bundling: A pricing strategy that combines multiple menu items into a grouping that is then sold at a price lower than that of the bundled items purchased separately.

The most common example of bundling is the combination meals (combo-meals) offered by quick-service restaurants. In many cases, these bundled meals consist of a sandwich, French fries, and a drink. Bundled meals, often promoted as "combo meals" or "value meals," are typically identified by a number (e.g., Number 6 or Number 7) or a single name [e.g., Everyday Value Meal (Arby's), Mix and Match (Burger King), and Cheeseburger Combo (McDonald's)] for ease of ordering.

Bundled menu offerings are carefully designed to encourage guests to buy all menu items included in the bundle, rather than to separately purchase only one or two of the items. Bundled meals are typically priced very competitively, so that a strong value perception is established in the guest's mind.

Find Out More

You have now been introduced to numerous factors that influence the prices foodservice operators charge for the items they sell. In the future, Leadership in Energy and Environmental Design (LEED) certification achieved by an operation may well constitute another such factor.

The LEED rating system developed by the U.S. Green Building Council (USGBC) evaluates facilities on several standards. The rating system considers sustainability, water use efficiency, energy usage, air quality, construction and materials, and innovation.

Increasingly, many consumers are willing to pay more to dine in LEED-certified operations. In addition, LEED-certified buildings are healthier for workers and for diners. The LEED certification creates benefits for foodservice operators, employees, and guests. It will likely continue to be of increasing importance to guests.

To learn more about LEED certification in the foodservice industry, enter "LEED-certified restaurant standards" in your favorite browser and view the results.

What Would You Do? 7.1

"We have to lower our prices because there is nothing else we can do!" said Ralph, director of operations for the seven-unit Boston Sub Shops. Boston's was known for its modestly priced, but very high-quality, sandwiches and soups.

Business and profits were good, but now Ralph and Rachel, who was Boston's Director of Marketing, were discussing the new $8.99 "Foot Long Deal" sandwich promotion just announced by their major competitor: an extremely large chain of sub shops that operates thousands of units nationally and internationally.

"They just decided to lower their prices to appeal to value-conscious customers," said Ralph.

"But how can they do that and still make money?" asked Rachel.

"There's always a less expensive variety of ham and cheese on the market," replied Ralph. "They use lower-quality ingredients than us, and we charge $10.99 for our foot-long sub. That wasn't bad when they sold theirs at $9.99. Our customers know we are worth the extra dollar. Now that they are running their special at $8.99 ... I don't know, but this might really hurt us," said Ralph, "What should we do?"

Assume you were Rachel. How do you think guests will respond to this competitor's new pricing strategy? What specific steps would you recommend to Ralph and Rachel that can help Boston's Subs address this new pricing/cost challenge?

Methods of Food and Beverage Pricing

Many factors impact how a commercial foodservice operation establishes its prices, and the methods used are often as varied as the operators who use the methods.

Menu item prices can be directly affected by one or more of the factors previously described. However, in most cases menu prices have historically been determined based on either an operation's costs or desired profit level per menu item. This makes sense when the operator's perspective of the price formula introduced earlier in this chapter is re-examined closely. That formula was presented as:

Item Cost + Desired Profit = Selling Price

When foodservice operators focus on item costs to establish prices, they recognize that menu items that cost more to produce must be sold at prices higher than lower-cost items. The actual prices charged in a foodservice operation are primarily determined by its owner, perhaps with the assistance of production staff. However, those responsible for foodservice accounting tasks should have a good understanding of the cost-based and profit-based approaches to pricing menu items.

Cost-based Pricing

When using cost-based pricing, a foodservice operator calculates the cost of the ingredients required to produce the menu item being sold. The cost of the menu item includes any menu item accompaniments. For example, when a foodservice operation sells a dinner entree including a salad and dinner rolls at one cost, the entree cost must also include that of the salad and dressing, and dinner rolls and butter included with the entree.

In the majority of cases, cost-based pricing is based on the idea that the cost of producing an item should be a predetermined percentage of the item's selling price. With this system, menu items with lower percentages of cost when compared to their selling prices are typically considered more desirable than menu items with higher percentage costs.

The formula used to compute a menu item's actual **food cost percentage** is:

Key Term

Food cost percentage: A ratio calculated by determining the food cost for a menu item and dividing that cost by the item's selling price.

$$\frac{\text{Item food cost}}{\text{Selling price}} = \text{Food cost percentage}$$

To illustrate, if it costs an operator $4.00 to purchase the ingredients needed for a menu item, and the item is sold for $20.00, the item's food cost percentage is calculated as:

$$\frac{\$4.00 \text{ item food cost}}{\$20.00 \text{ selling price}} = 0.20 \text{ or } 20\%$$

When operators utilize food cost percentage as a major factor in pricing menu items, the cost of producing one portion of the menu item must be accurately determined. When the item sold is a single portion item (e.g., one New York strip steak), the cost of producing one portion is relatively straightforward.

However, when menu items are produced in multiple portions, for example when a pan of lasagna containing 12 portions is prepared, operators must calculate their **portion cost** based on the standardized recipe (see Chapter 6) used to produce the menu item.

Key Term

Portion cost: The product cost required to produce one serving of a menu item.

The use of standardized recipes is critical for operators utilizing product cost as a primary determinant of menu prices. The cost of producing a menu item must be known, and it must be consistently the same if selling prices are based on product costs. When ingredient costs change, the increased cost must also be used to calculate the new portion costs resulting from standardized recipes now containing higher cost ingredients. These new portion costs may (or may not) cause an operator to change menu prices, but the actual costs must be known.

While for-profit foodservice operations are concerned with food cost percentages, non-profit operations are typically more interested in the cost to serve each guest.

Examples of non-profit operations in which cost per meal is important to operators include military bases, prisons, hospitals, senior living facilities, schools and colleges, and large business organizations that offer foodservices.

Whether the individuals served are soldiers, inmates, patients, residents, students, or employees, calculating an operation's cost per meal is easy because it uses a variation of the basic food cost percentage formula:

$$\frac{\text{Cost of Food Sold}}{\text{Total Meals Served}} = \text{Cost Per Meal}$$

For example, assume a non-profit operation incurred $65,000 in cost of food sold during an accounting period. In the same accounting period, the operation served 10,000 meals. To calculate this operation's cost per meal, the cost per meal formula is applied.

In this example, it would be:

$$\frac{\$65,000 \text{ Cost of Food Sold}}{10,000 \text{ Meals Served}} = \$6.50 \text{ Cost Per Meal}$$

Whether managers are most interested in their cost of food percentage or in their cost per meal served, it is essential that they first accurately calculate the cost of food sold. When foodservice operators have established a target food (or beverage) cost percentage, they can use it to set their menu prices.

For example, if a menu item costs $8.00 to prepare, and an operation's desired cost percentage is 40%, (the food cost percentage in the approved operating budget for the time the menu is used), the following formula can determine the item's menu price:

$$\frac{\text{Food Cost of Menu Item}}{\text{Desired Food Cost}\%} = \text{Selling Price}$$

In this example:

$$\frac{\$8.00 \text{ Food Cost of Menu Item}}{40\% \text{ Desired Food Cost}} = \$20.00 \text{ Selling Price}$$

Another method of calculating selling prices based on predetermined product cost percentage goals uses a pricing factor (multiplier) assigned to each potentially desired food or beverage cost percentage. This factor when multiplied by an item's portion cost indicates a selling price that yields the desired product cost percentage. Some of the most commonly used pricing factors are presented in Figure 7.2.

The above pricing factor method of establishing menu prices is easy to use. For example, if an operator wants a 25% product cost, and a menu item has a food cost of $4.50, the following pricing formula would be used:

Item Food Cost × Pricing Factor = Selling Price

In this example, that would be:

$4.50 Item Food Cost × 4.0 Pricing Factor = $18.00 Selling Price

The two methods to determine a proposed selling price based on product cost percentage yield identical results. With either approach, an item's selling price is determined with the goal of achieving a specified food or beverage cost percentage for each item sold.

Desired Product Cost %	Pricing Factor
20	5.000
23	4.348
25	4.000
28	3.571
30	3.333
$33^{1}/_{3}$	3.000
35	2.857
38	2.632
40	2.500
43	2.326
45	2.222

Figure 7.2 Pricing Factor Table

Technology at Work

Foodservice professionals will have no difficulty identifying numerous publications detailing a variety of cost-based menu pricing techniques and strategies.

To find a list of currently released publications, go to the Amazon website, select "Books" from the pull-down menu, and then enter either *Menu Pricing* or *Food Cost Control* to review the most recently published works related to the determination of food and beverage menu prices.

One of the best on the market is *Food and Beverage Cost Control*, published by John Wiley, and written by Dr. Lea Dopson and Dr. David Hayes. When you are on the Amazon site, review the index of the latest edition of this popular text.

Contribution Margin-based Pricing

Some commercial foodservice operators use a more profit-based pricing method to establish menu item selling prices. These operators set their prices based on each of their menu item's contribution margin (CM) (see Chapter 6).

CM for a single menu item is defined as the amount of money that remains after the product cost of the menu item is subtracted from the item's selling price. It is, then, the amount that a menu item "contributes" to pay for labor and all other expenses, and to contribute to the operator's profit margin.

To illustrate CM pricing, assume a menu item sells for $18.75 and the food cost to produce the item is $7.00. In this example, the CM for the menu item is calculated as:

Selling Price − Item Food Cost = Contribution Margin (CM)

or

$18.75 Selling Price − $7.00 Item Food Cost = $11.75 CM

When this approach is used, the formula for determining a menu item's selling price is:

Item Food Cost + Contribution Margin (CM) Desired = Selling Price

When using the CM approach to establish selling prices, operators develop different CM targets for various menu items or groups of items. For example, in an operation where items are priced separately, entrées might be priced with a CM of $10.50 each, desserts with a CM of $5.25 each, and non-alcoholic drinks with a CM of $3.75.

To apply the CM method of pricing, foodservice operators use a two-step process.

Step 1: Determine average contribution margin required per item.

Step 2: Add the contribution margin required to the item's product cost.

Step 1: Operators determine the average CM they require based on the number of items to be sold or on the number of guests to be served. The process used for each approach is identical.

For example, to calculate CM based on the number of items to be sold, operators add their non-food operating costs to the amount of profit they desire, and then divide the result by the number of items expected to be sold:

$$\frac{\text{Non-Food Costs} + \text{Profit Desired}}{\text{Number of Items to Be Sold}} = \text{CM Desired Per Item}$$

To calculate CM based on the number of guests to be served, operators divide all of their non-food operating costs, plus the amount of profit they desire, by the number of expected guests:

$$\frac{\text{Non-Food Costs} + \text{Profit Desired}}{\text{Number of Guests to Be Served}} = \text{CM Desired Per Guest Served}$$

For example, if an operator's budgeted non-food operating costs for an accounting period are $125,000, desired profit is $15,000, and the number of items estimated to be sold is 25,000, the operator's desired average CM per item would be calculated as:

$$\frac{\$125,000\,(\text{Non-Food Costs}) + \$15,000\,(\text{Profit Desired})}{25,000\,(\text{Number of Items to Be Sold})} = \$5.60\,\text{CM Desired Per Item}$$

Step 2: Operators complete this step by adding their desired CM per item (or guest) to the cost of preparing a menu item. For example, if (example above) an operator's desired average CM per item is $5.60 and a specific menu item's food cost is $3.40, the item's selling price would be calculated as:

$$\$5.60\,\text{CM Desired} + \$3.40\,\text{Item Food Cost} = \$9.00\,\text{Selling Price}$$

The CM method of pricing is popular because it is easy to use, and it helps ensure each menu item sold contributes to an operation's profits (recall that desired profit is included with non-food costs) when the calculation is made. When using CM to set menu prices, the prices charged for menu items vary only due to variations in product cost. When managers have accurate budget information about their non-food costs (taken from previous income statements) and realistic profit expectations, the use of the CM method of pricing can be very effective.

Operators who utilize the CM approach to pricing do so believing the average CM per item sold is a more important consideration in pricing decisions than is the product cost percentage. The debate over the "best" pricing method for food and beverage products is likely to continue.

All operators must view pricing as an important process and its goal is to consider a desirable price/value relationship for guests. In the final analysis, the customer eventually determines what an operation's sales will be for each menu item. Experienced operators know that sensitivity to required profit and to guests' needs, wants, and desires are the most critical components of an effective pricing strategy.

Evaluation of Pricing Efforts

Regardless of which method is used to establish selling prices, foodservice operators should regularly evaluate the results of their pricing efforts, and many operations evaluate menu items based on two key characteristics: popularity and profitability.

Menu engineering is a term popularly used to describe one method that addresses these two variables.

Menu Engineering

Operators using the menu engineering process attempt to produce a menu that maximizes the menu's overall CM (defined earlier as the amount operators have available to pay for labor and all other expenses and to contribute to profit). As a result, operators should be keenly focused on their menu's overall CM.

Key Term

Menu engineering: A system used to evaluate menu pricing and design by categorizing each menu item into one of four categories based on its profitability and popularity.

To use menu engineering, operators must sort their menu items by two variables:

1) Popularity (number of each item sold)
2) Weighted contribution margin

Calculating Popularity (Number Sold)

To calculate the average popularity (number sold) of a menu item, operators must determine the average number sold:

$$\frac{\text{Total Number of Menu Items Sold}}{\text{Number of Menu Items Available}} = \text{Average Number Sold}$$

For example, if an operator sold 5,000 entrees in a specific time period, and the operator's menu lists 10 different entree choices, the average popularity of the entrees sold in the time period would be calculated as:

$$\frac{5,000 \text{ Menu Items Sold}}{10 \text{ Menu Items Available}} = 500 \text{ Menu Item Average Popularity (Number Sold)}$$

When using menu engineering and applying the 500 average popularity, any menu item that sold *more* than 500 times during the analysis period is classified as "High" in popularity, and menu items selling *less* than 500 times would be classified as "Low" in popularity.

Calculating Weighted Contribution Margin

To continue the menu engineering process, operators must also define the weighted average CM of their menu items. Some operators confuse averages (means) with weighted averages. However, the distinction between the two is important.

To use a simple example, assume an operator collected the following data and wanted to calculate the average size of the sale made in their operation over the three-day reporting period.

Week Day	Guests Served	Total Sales	Average Sale
Monday	50	$ 500	$10.00
Tuesday	150	$1,650	$11.00
Wednesday	250	$3,000	$12.00
Total/Average	450	$5,150	?

In this example, to calculate the size of the "average" sale on Monday through Wednesday, one should NOT use the unweighted formula typically used to calculate a mean (average). That *unweighted formula* is:

$$\frac{\$10.00 + \$11.00 + \$12.00}{3 \text{ days}} = \$11.00 \text{ per day average}$$

In fact, what the operator really wants to learn when calculating the average sales for the three days is "How much did the average guest spend in my operation from Monday through Wednesday?"

The number of guests served each day varied, so the operator must determine a *weighted* average sale formula as shown below:

$$\frac{\$5,150 \text{ (Total Sales During All 3 Days)}}{450 \text{ (Total Guests Served in All 3 Days)}} = \$11.44 \text{ Weighted Average Sale Per Guest}$$

Note that the operator's average sale size resulting from using unweighted and weighted average formulas in this example differ.

Returning to menu engineering, to calculate the average **weighted contribution margin** for their menu items, operators must first calculate the total contribution margin generated by all items sold, and then divide by the number of items sold as shown in Figure 7.3.

Column A in Figure 7.3 lists the name of individual menu items on the menu. Column B lists the total number of each menu item that was sold. Note that the sales (popularity) of the items vary from a low of 190 sold (Item 10) to a high of 1,050 sold (Item 4). Column C lists the individual CM of each of the 10 items offered for sale, and column D lists the total item CM generated by each menu item.

Key Term

Weighted contribution margin: Weighted contribution margin: The contribution margin provided by all menu items divided by the total number of items sold. Weighted contribution margin is calculated as:

$$\frac{\text{Total Contribution Margin of All Items Sold}}{\text{Total Number of Items Sold}}$$
$$= \text{Weighted Contribution Margin}$$

Column A	Column B	Column C	Column D
Menu Item	Total Number Sold	Single Item Contribution Margin	Total Item Contribution Margin
1	250	$14.50	$3,625
2	250	$7.50	$1,875
3	525	$12.50	$6,563
4	1,050	$17.25	$18,113
5	510	$7.00	$3,570
6	625	$13.50	$8,438
7	400	$12.75	$5,100
8	825	$10.50	$8,663
9	375	$8.25	$3,094
10	190	$16.50	$3,135
Total	5,000	$120.25	$62,174
Average (Mean)	500	**$12.03**	
Weighted contribution margin			**$12.43**

Figure 7.3 Total Item Contribution Margin Worksheet for 10 Item Menu (rounded to the nearest dollar)

		Popularity	
		Low	High
Contribution Margin	High	High contribution margin Low popularity PUZZLE	High contribution margin High popularity STAR
	Low	Low contribution margin Low popularity DOG	Low contribution margin High popularity PLOW HORSE

Figure 7.4 Menu Engineering Matrix

The value in Column D is calculated by multiplying the value in Column B times the value in Column C. In this example, the average number sold is 500, and the average weighted CM is $12.43 ($62,174 total item CM/5,000 sold = $12.43).

After an operator has calculated the popularity and weighted contribution margin of the items listed on the menu, the items are sorted into a 2 × 2 menu engineering matrix containing four squares as shown in Figure 7.4.

Figure 7.4 shows that menu items with sales above the average level of popularity (500 sold in this example) are "High" in popularity, and items that sold less than 500 times are "Low" in popularity. Similarly, those menu items whose contribution margins are above the weighted contribution margin average ($12.43 in this example) are "High" in contribution margin, and items with a lower contribution margin are "Low" in contribution margin.

Many users of menu engineering name the items contained in the four squares for ease of remembering the characteristics of each item. These commonly used names (Puzzles, Stars, Dogs, and Plow Horses) are also shown in Figure 7.4.

Figure 7.5 shows where each of the 10 example menu items listed in Figure 7.3 would be located.

		Popularity	
		Low	High
Contribution Margin	High	PUZZLE Menu items 1, 7, and 10	STAR Menu items 3, 4, and 6
	Low	DOG Menu items 2 and 9	PLOW HORSE Menu items 5 and 8

Figure 7.5 Menu Engineering Results

Item	Characteristics	Problem	Marketing Strategy
Puzzle	High contribution margin, Low popularity	Marginal due to lack of sales	a) Relocate on menu for greater visibility. b) Consider reducing the selling price.
Star	High contribution margin, High popularity	None	a) Promote well. b) Increase prominence on the menu.
Dog	Low contribution margin, Low popularity	Marginal due to low contribution margin and lack of sales	a) Remove from the menu. b) Consider offering as a special occasionally, but at a higher menu price.
Plow Horse	Low contribution margin, High popularity	Marginal due to low contribution margin	a) Increase the menu price. b) Reduce prominence on the menu. c) Consider reducing portion size.

Figure 7.6 Potential Menu Modifications

Menu Modifications

Operators should regularly analyze menus to make needed modifications and improvements. When using menu engineering, each menu item that fell within the four squares requires a special marketing strategy. Examples of these suggested menu modification strategies resulting from menu engineering are summarized in Figure 7.6.

Technology at Work

A foodservice operation's menu is much more than a list of foods and beverages. It can be a powerful marketing tool to familiarize customers with a brand. It can also get them excited about the unique items an operation offers for sale.

Regularly evaluating individual menu items for popularity and profitability is an important marketing task because it helps identify both profit producing items and those that perform poorly. With this information known, menus can be modified to optimize sales and profits.

Fortunately, there are several useful software programs available that help operators perform menu engineering analysis. To review the features and costs of such programs, enter "menu engineering software" in your favorite search engine and view the results.

One reason to perform menu engineering analysis is to identify items whose prices must or can be increased to enhance an operation's profitability. Some operators are hesitant to raise prices fearing customers will react negatively. Experienced foodservice operators, however, know that price is not the only determining factor when guests decide where to spend their dining-out dollars.

Quality of customer service, cleanliness, staff friendliness, and uniqueness of menu items offered are often *more* important than price. The best operators ensure all these aspects of their operations meet or exceed their guests' expectations. Then, price increases acceptable to guests and that help ensure an operation's long-term profitability can be implemented.

This chapter has indicated that foodservice operators base their prices on production costs and desired profit. Properly controlling the cost of menu items as they are prepared and served requires specific systems and skills. In fact, controlling food and beverage production costs are so important that this is the sole topic of the next chapter.

What Would You Do? 7.2

"$39.95 – that's over ten dollars more than we charged for it yesterday!" said Shawn, the Dining Room manager at Chez Franco's restaurant.

Shawn was discussing the day's dinner menu with Aimée, the restaurant's executive chef. Aimée had just shown Shawn the daily menu insert his service staff would use that night. On the night's new menu, he noticed that the price of Lemon Red Snapper with Herb Butter, one of the operation's most popular dishes, had increased overnight. Yesterday it sold for $27.95. Today Aimée had priced it at $39.95.

"Tell me about it," replied Aimée, "our seafood supplier really raised the price on our latest delivery. My snapper cost is now up by almost $11.00 per pound. That's over $4.00 a portion. The supplier said there was a snapper shortage, and he wasn't sure how long it would last. With the new cost of snapper, I needed this price increase to keep our food cost percentage in line.

Shawn wasn't sure his servers or the guests they would serve that night would be very happy with Aimée's pricing decision. The snapper was a very popular item, and that meant tonight lots of customers would likely notice the price increase and comment about it!

Assume you were a server at Chez Franco's on the night this new menu price was initiated. Unless you were instructed otherwise, how would you likely respond to a returning guest who questioned the significant price increase on the red snapper item? What do you think Shawn should tell his servers to say in response to guests' anticipated reactions to this menu item's price increase?

Key Terms

Price (noun)	Wagyu beef	Food cost percentage
Price (verb)	Target market	Portion cost
Profit margin	Third-party delivery	Menu engineering
Value	Restaurant row	Weighted contribution
Consumer rationality	Bundling	margin

Operator's 10-Point Tactics for Success Checklist

Evaluate your need for, and the current status of, each of the following operational tactics. For those tactics you think are important, but not yet in place, develop an action plan for its implementation including who will be responsible for the tactic's completion and the target date by which it should be completed.

				If Not Done	
Tactic	Don't Agree (Not Done)	Agree (Done)	Agree (Not Done)	Who Is Responsible?	Target Completion Date
1) Operator understands the importance of price in the profitable operation of a foodservice business.	——	——	——		
2) Operator has carefully considered the difference between an operator's view of price and their guests' view of price.	——	——	——		
3) Operator has considered the impact of economic conditions, local competition, level of service, and guest type when establishing menu prices.	——	——	——		
4) Operator has considered the impact of portion size, delivery style, meal period, and location when establishing menu prices.	——	——	——		
5) Operator understands the concept of bundling when establishing menu prices.	——	——	——		

(Continued)

Tactic	Don't Agree (Not Done)	Agree (Done)	Agree (Not Done)	If Not Done	
				Who Is Responsible?	Target Completion Date
6) Operator has considered the value of utilizing a product cost-based approach when establishing menu prices.	___	___	___		
7) Operator has considered the value of utilizing a contribution margin approach when establishing menu prices.	___	___	___		
8) Operator understands the importance of analyzing the menu as a means of better understanding guests' purchasing preferences.	___	___	___		
9) Operator understands how to use menu engineering to analyze their menus.	___	___	___		
10) Operator recognizes the importance of carefully implementing any menu price adjustments to minimize guest dissatisfaction.	___	___	___		

8

Food and Beverage Cost Control

Operator's Brief

In this chapter, you will learn that proper control of product costs is essential to foodservice success. When product costs are professionally managed, guest satisfaction is optimized, and an operation can better achieve its targeted profit goals.

Two important tools used by foodservice operators to control product costs are sales forecasts and standardized recipes. Accurate sales forecasts help foodservice operators estimate how many guests will be served and what they will buy. Standardized recipes indicate how much product should be purchased and produced to meet these estimates.

When foodservice operators understand what guests will purchase during their visits, they can place their orders for necessary food and beverage products. When necessary items are delivered, great care is needed to ensure products received match those management intended to buy. This process is achieved with purchase specifications (specs), formal purchase orders (POs), and use of proper receiving procedures.

Received products must be properly and securely stored until used. Careful control of stored products minimize product loss from waste, spoilage, and theft. Attention to details is critical as products are produced and served in intended portion sizes.

(Continued)

Even when proper purchasing, receiving, storage, and production controls are used, some operators may find that planned product cost percentages exceed their targeted levels. Operators can then choose from numerous strategies to help reduce product cost percentages to levels that assist them in achieving their targeted profit goals.

CHAPTER OUTLINE

The Importance of Food and Beverage Cost Control
    Sales Forecasts
    Standardized Recipes
Purchasing and Receiving Products
    Purchasing Food and Beverage Products
    Receiving Products
Managing Inventory and Production
    Placing Products in Storage
    Maintaining Product Security
    Managing Production
Controlling the Cost of Sales Percentage
    Controlling Food Costs
    Controlling Beverage Costs
    Optimizing Cost of Sales

The Importance of Food and Beverage Cost Control

A foodservice operator's primary responsibility is to deliver quality products and services to guests at prices mutually agreeable to both parties. In addition, product and service quality must enable guests to believe excellent value was received for the money they spent. When quality and value are met, an operation will prosper. If, however, management focuses more on reducing costs than providing value, problems will inevitably occur. This is why operators must focus on "controlling" costs rather than on minimizing or eliminating costs.

Every foodservice operator must know, understand, and practice the basic profit formula introduced in Chapter 3:

Revenue – Expenses = Profit

Figure 8.1 summarizes how operators generate profits by using the cash (revenue) received from the sale of their products and services.

When the process shown in Figure 8.1 includes careful control of product purchasing and production costs, guest satisfaction and profits can be optimized. Guests directly benefit because an operator's prices (see Chapter 7) support the delivery of a specific quantity and quality of product for the price paid. Successful

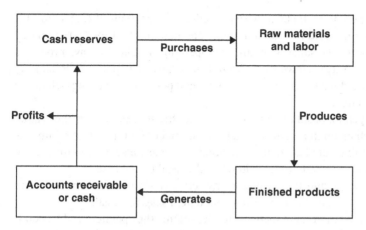

Figure 8.1 Foodservice Business Flowchart

foodservice operators carefully purchase products to ensure intended quality can be obtained, they prepare products that optimize quality, and they also deliver the products in a way that ensures guests receive good value.

A foodservice operation directly benefits when product costs are properly controlled. Improper purchasing, inadequate inventory control, and production errors contribute to excessive costs that reduce profits. Two major tools operators utilize to control product costs are sales forecasts and standardized recipes.

Sales Forecasts

A properly produced **sales forecast** estimates the number of guests a foodservice operation will serve during a specific time, and what those guests will order. For example, assume that, based on prior sales, an operator estimates 300 guests will visit the operation for lunch next Monday.

Key Term

Sales forecast: An estimate of the number of guests to be served and the menu items they will purchase within a specific time period.

Assume further that the operation sells only three entrée items: roast chicken, roast pork, and roast beef. A question the operator must consider is, "How many servings of each item should be produced so we do not run out of any item nor have too many of each item remaining after the lunch period?"

If the operation ran out of one of the three menu items, guests who wanted that item would likely be upset. They might even leave the operation (perhaps taking dining companions with them who would have ordered other items the operation produced). Producing too much of any one item would result in unsold items that would cause product costs to rise unnecessarily unless these remaining items could later be sold at full price.

In this situation, it would be unwise to produce 300 portions of each item. If it did, the operation would not run out of any one item (each of the 300 estimated guests could order the same item, and the operation would still have produced a sufficient quantity). However, it would also have 600 portions remaining (900 portions produced − 300 portions sold = 600 portions remaining) when the lunch period ended.

What this operator would like to do is instruct the staff to prepare the "right" amount of each menu item. This would be the number of servings that minimizes chances of running out of an item before lunch is over and also reduces the chance of having an excessive amount unsold when the meal period ends.

The answer to the question of how many servings of roast chicken, pork, and beef should be prepared lies in accurate menu item sales forecasting.

Returning to the three-item menu example, assume the operator used its point-of-sale (POS) system to record last week's sales of menu items on a form similar to the one presented in Figure 8.2.

A review of the Figure 8.2 data indicates an estimated 300 guests for next Monday makes good sense because the weekly sales total last week of 1,500 guests served averages 300 guests per day (1,500 guests served/5 days = 300 guests served per day).

In this example, last week the operation sold an average of 73 roast chicken (365 sold/5 days = 73 per day), 115 roast pork (573 sold/5 days = 115 per day), and 112 roast beef (562 sold/5 days = 112 per day) portions per day.

Assume an operator knows the average number of people selecting a given menu item and they know the total number of guests who made the selections, they can compute each menu item's **popularity index:** the proportion of total guests choosing a given menu item from a list of alternative menu items.

Key Term

popularity index: The proportion of guests choosing a specific menu item from a list of alternative menu items. The formula for a popularity index is:

$$\frac{\text{Total Number of a Specific Menu Item Sold}}{\text{Total Number of All Menu Items Sold}}$$
$$= \text{Popularity Index}$$

Date: 1/2–16 Menu Items Sold

Menu Item	Mon	Tues	Wed	Thurs	Fri	Week's Total	Average Daily Sales
Roast Chicken	70	72	61	85	77	365	73
Roast Pork	110	108	144	109	102	573	115
Roast Beef	100	140	95	121	106	562	112
Total	280	320	300	315	285	1,500	300

Figure 8.2 Menu Item Sales History

In this example, the operator could improve the estimate about the quantity of each item to prepare by using historical sales to help guide production planning. If it is assumed that future guests will select menu items in a manner similar to that of past guests, the information can improve sales forecasts using the following formula:

$$\frac{\text{Total Number of a Specific Menu Item Sold}}{\text{Total Number of All Menu Items Sold}} = \text{Popularity Index}$$

The popularity index for roast chicken sold last week would be 24.3% (365 roast chickens sold/1,500 total guests = 0.243, or 24.3%). Similarly, 38.2% (573 roast pork sold/1,500 total guests = 38.2%) of guests preferred roast pork, and 37.5% (562 roast beefs sold/1,500 total guests = 37.5%) selected roast beef.

In most cases, when an operator knows the number of menu items future guests will select, they are better prepared to make good decisions about the quantity of each item to produce. The basic formula for individual menu item forecasting based on the item's sales history is:

Number of Guests Expected × Item Popularity Index = Predicted Number to Be Sold

In this example, Figure 8.3 illustrates the operator's best estimate of what their 300 guests are likely to order.

The predicted number to be sold is the quantity of a specific food or beverage item likely served given a reasonable estimate of the total number of guests expected to be served. When operators know which items guests are likely to select, they can determine how much of each menu item their production staff should prepare or have ready to serve.

It is important to recognize that, while sales forecasts cannot precisely estimate the number of guests who may arrive on any given day (nor the items they will buy!), use of these forecasts is essential to the control of food and beverage production costs.

Menu Item	Guest Forecast	Popularity Index	Predicted Number to Be Sold
Roast Chicken	300	0.243	73
Roast Pork	300	0.382	115
Roast Beef	300	0.375	112
Total			300

Figure 8.3 Forecasted Item Sales

Standardized Recipes

Although the menu determines what is to be sold and at what price, the standardized recipe (see Chapter 6) controls the quantity and quality of what a kitchen or bar will produce.

A standardized recipe details the procedures to prepare and serve all of an operation's food or beverage items. A standardized recipe ensures that each time guests order a menu item they receive exactly what management intended them to receive. A high-quality standardized recipe will contain the following information:

1) Menu item name
2) Total recipe yield (number of portions produced)
3) Portion size
4) Ingredient list
5) Preparation/method section
6) Cooking time and temperature
7) Special instructions, if necessary
8) Recipe cost (optional)

Figure 8.4 shows a sample standardized recipe for a roast chicken item. This standardized recipe represents the quality and quantity the operation wishes its

Roast Chicken

Special Instructions: Serve with Crabapple Garnish (see Crabapple Garnish Standardized Recipe)

Recipe Yield: 48 portions
Portion Size: ¼ chicken

Serve on 10-in. plate.

Ingredients	Amount	Method
Chicken Quarters (twelve 3–3½-lb. chickens)	48 ea.	Step 1. Wash chicken; check for pinfeathers; tray on 24 inch × 20 inch baking sheets.
Butter (melted)	1 lb. 4 oz.	Step 2. Clarify butter; brush liberally on chicken quarters; combine all seasonings; mix well; sprinkle all over chicken quarters.
Salt	¼ C	
Pepper	2 T	
Paprika	3 T	
Poultry Seasoning	2 t	
Ginger	1½ t	
Garlic Powder	1 T	Step 3. Roast in oven at 325 °F for 1.5 hours, or to an internal temperature of at least 165 °F.

Figure 8.4 Standardized Recipe: Roast Chicken

guests to receive. If it is followed carefully each time, guests will always receive the quality and value the operation intended.

The best foodservice operators create and require the use of standardized recipes because:

1) Accurate purchasing is impossible without the existence and use of standardized recipes.
2) Dietary concerns require some foodservice operators to know the exact ingredients and correct amount of nutrients in each serving of a menu item.
3) Accuracy in menu laws requires that foodservice operators inform guests about the type and amount of ingredients in their recipes.
4) The responsible service of alcohol is impossible without adhering to appropriate portion sizes in standardized drink recipes.
5) Accurate recipe costing and menu pricing is impossible without standardized recipes.
6) Matching food and beverages used to menu items sold is impossible without standardized recipes.
7) New employees can be trained faster and better with standardized recipes.
8) The computerization of a foodservice operation is impossible unless the elements of standardized recipes are in place. The advantages of advanced technological tools available to the operation are restricted or even eliminated if standardized recipes are not used.

Experienced foodservice operators agree that use of standardized recipes is essential to any serious effort to produce consistent, high-quality food and beverage products at a specific cost.

Technology at Work

Several good companies offer software programs that ease standardized recipe development and needed measurement conversion. Note: Many recipes use metric measurements rather than imperial measurements. Software programs help operators manage their standard recipes files, and they can also:

✓ Calculate recipe costs
✓ Help create new recipes
✓ Plan menus
✓ Conduct nutritional analysis of recipes
✓ Track critical allergen information
✓ Monitor product inventory levels in real time

To examine some of these innovative and increasingly essential cost control tools, enter "standardized recipe development software" in your favorite search engine and review the results.

Purchasing and Receiving Products

Sales forecasts and the use of standardized recipes allow operators to know exactly what must be purchased to ensure they offer the menu items their guests want to purchase. Entire books have been written about foodservice purchasing and purchasing systems. However, proper foodservice purchasing is essentially a two-part process:

1) Purchasing food and beverage products
2) Receiving products

Purchasing Food and Beverage Products

The question of what food or beverage product should be purchased is answered by the use of a **product specification** ("spec") for each needed item.

A product specification is the way foodservice operators precisely communicate their product needs with vendors (suppliers), so their operations receive the exact item they want every time it is ordered.

Key Term

Product specification (spec): A detailed description of a recipe ingredient or complete menu item to be served.

A product specification determines neither the best product to buy nor the product that costs the least. Rather, it is the product determined to be the *most appropriate* for its intended use in terms of both quality and cost.

A product specification that is not sufficiently detailed can create problems because the needed level of item quality may not be delivered. However, if product specifications are written too tight (they are overly and unnecessarily specific), then too few vendors may be able to supply the products needed. The result: excessively high costs from the few vendors who can deliver the items.

Figure 8.5 is an example of a professionally prepared product specification for the bacon used by a foodservice operation for bacon, lettuce, and tomato (BLT) sandwiches.

A properly prepared product specification includes the following information:

1) Product name or specification number
2) Pricing unit
3) Standard or grade
4) Weight range/size
5) Processing and/or packaging
6) Intended use

Product Name:	Bacon, sliced	Spec #: 117
Pricing Unit:	lb.	
Standard/Grade:	Select No. 1	
	Moderately thick slice	
	Oscar Mayer item 2040 or equal	
Weight Range:	14–16 slices per lb.	
Packaging:	2/10 lb. Cryovac packed	
Container Size:	Not to exceed 20 lb.	
Intended Use:	Bacon, lettuce, and tomato sandwiches	
Other Information:	Flat packed on oven-proofed paper	
	Never frozen	
Product Yield:	60% Yield (usable amount)	

Figure 8.5 Sample Product Specification: Bacon (sliced)

Product Name (Specification Number)

Product names are not always self-explanatory. For example, chickpeas and garbanzo beans are the same item, but they may be packaged under either name. The same is true for sweet potatoes and yams (although they are technically different items). Similarly, large shrimp and prawns may mean the same product to some vendors but not to others. Therefore, a product's name on a specification must be detailed enough to clearly and precisely identify the item an operator wants to buy.

When developing product specifications, many operators find it helpful to assign a number and a name to the item they wish to buy. This can be useful when, for example, many forms of the same ingredient or menu item are purchased.

To illustrate, consider a deli-restaurant that may use 5–10 different types of bread depending on its intended use. Each type of bread may be given a specific name and spec number. The same may be true with other needed items such as cheese that is available in a variety of forms (e.g., block, sliced, or shredded) and types (e.g., Colby, Cheddar, or Swiss). Note that, in Figure 8.5, this bacon is given a name and a specification number (Spec #117).

Pricing Unit

A product's pricing unit may be pounds, quarts, gallons, cases, or other commonly used measurement unit in which the item is sold. Many pricing units such as pounds and quarts are easily understood. However, some pricing units are less well known. Parsley, for example, is typically sold in the United States by the bunch, and grapes are sold by the lug. Other commonly used pricing units include cartons, trays, bushels, and flats. To avoid confusion, operators should insist that their vendors supply them with definitions of each pricing unit upon which they base their selling prices.

Standard or Grade

Many food items are sold with varying degrees of quality. The U.S. Department of Agriculture (USDA), the Bureau of Fisheries, and the Food and Drug Administration have developed standards (grades) for many food items. In addition, grading programs are in place for many commonly used foodservice items. Trade groups such as the National Association of Meat Purveyors publish item descriptions for many of these products. Consumers also are aware of many of these distinctions. One good example of this is the USDA's grades for beef used in most foodservice operations. In descending order of quality, these are:

✓ Prime
✓ Choice
✓ Select
✓ Standard (Commercial)

In some cases, operators may prefer a specification identifying a particular product's brand name (e.g., Minor's beef base or A-1 Steak Sauce) or a specific point of origin (e.g., Maine lobster or Idaho potatoes) as an alternative or supplement to a product standard or grade.

Weight Range/Size

Weight range or size is important when purchasing some meats, and some fish, poultry, fruits, and vegetables. In the standardized recipe example of roast chicken (see Figure 8.4), the chicken quarters should be from chickens weighing 3.0–3.50 pounds. Note: The weight range would make a big difference in portion cost and cooking times compared to, for example, chickens weighing more or less than this weight.

When products require a specific trim or maximum fat covering, this should also be designated. For example, a 10-ounce strip steak with maximum tail of 1 inch and fat covering of 1/2 inch. Four-ounce hamburger patties, 16-ounce T-bone steaks, and ¼-pound hot dogs are examples of items that require an exact size rather than a weight range.

Count is a purchasing term used to designate product size. For example, 16- to 20-count

Key Term

Count (product): A number used to designate product size or quantity when purchasing food items.

shrimp means that, for this size shrimp, 16–20 individual shrimp are in one pound. Similarly, 30- to 40-count shrimp means there are 30–40 of this sized shrimp in one pound. In addition to seafoods such as shrimp and scallops, many fruits and vegetables are sold by count. In general, the larger the count, the smaller the size of the individual food items.

Processing and/or Packaging

Processing and packaging refer to a product's state when purchased. Peaches, for example, may be purchased fresh, dried, canned, or frozen. Each form specifies an appropriate price when it is purchased.

Food can come packed in numerous forms and styles including slab packed, layered cell packed, fiberboard divided, shrink-wrap packed, individually wrapped, and bulk packed. The scope of this book does not detail all of the many varieties of food processing and packing styles available, but it is important to know about them. Vendors chosen by the operator should provide information about all processing and packaging alternatives offered for the products they sell.

Find Out More

"Farm to Fork" (or "Farm to Table") is a term of importance to foodservice operators and their guests. The term refers to the path food follows from those who grow or raise it to those who will prepare and serve it. Ideally, this path is short to maximize freshness, minimize health risks, and be environmentally friendly. For that reason, many foodservice operators seek out and buy locally grown foods whenever possible.

In addition to freshness, locally grown foods are most often good for the environment. Less storing, shipping, and packaging results in less energy to transport the items and less solid waste from excessive packing materials. Reduced transportation and packaging costs can also translate into lower prices charged to foodservice operators.

Locally grown and organic products also help support the same local economies from which many foodservice operators draw their customers. As a result, buying locally is an option that should be thoroughly examined for benefits to their communities, the environment, the health of customers, and profit levels.

To learn more about this popular purchasing approach, enter "Farm to Fork trends," or "Farm to Table trends," in your favorite browser and view the results.

Intended Use

In foodservice operations, different types of the same item are often used in a variety of ways. Consider the operator who uses strawberries. Perfectly colored and shaped large berries are best for chocolate-dipped strawberries served on a buffet table. Less-than-perfect berries, however, may cost less and be perfectly acceptable for sliced strawberries used to top strawberry shortcake. Frozen berries may make a good choice for a baked strawberry pie and may even be of lower cost.

Breads, dairy, apples, and other fruits are additional examples of foods that are available in several forms. The best form of a food product is not necessarily the most expensive form. Rather, it is the form that is best for a product's intended use.

Regardless of the food or beverage items to be purchased, all operators should use a **purchase order (PO)** to formally request delivery of necessary items.

A PO may include a variety of product information but must always include the quantity ordered and the price quoted by the supplier. When ordering products, operators may choose several ways to tell suppliers what they want to buy. These may include ordering via the Internet, by telephone, or in person. Regardless of the communication method utilized, it is critical that there is a written PO because it serves as the formal record of what will be purchased.

POs can be simple or complex but should include the following information:

1) Vendor contact information
2) Purchase order number
3) Date ordered
4) Delivery date requested
5) Name of person who placed order
6) Names of ordered items
7) Item specification #, if appropriate
8) Quantity ordered
9) Quoted price (per unit)
10) Total PO cost
11) Delivery instructions (if applicable)
12) Comments (optional)

Figure 8.6 shows a sample PO for various produce items ordered from Scooters Produce by Joshua Jenn, the operator of the ABC restaurant.

Key Term

Purchase order (PO): A detailed listing of products a buyer is requesting from a vendor.

Receiving Products

After a PO is submitted, the ordered items must be properly received when delivered. A specially designated receiving clerk (in a large operation) may perform the receiving function, but in smaller operations a manager, supervisor, or a designated staff member likely is responsible.

ABC Restaurant Purchase Order

Vendor: Scooter's Produce	**Purchase Order #:** 256
Vendor's Address: 123 Anywhere Street, Any City, Any State	**Delivery Date:** 1/18
Vendor's Telephone #: 1-800-999-0000	**Vendor's E-Mail:** scootersproduce@isp.org

Item Purchased	Spec #	Quantity Ordered	Quoted Price	Extended Price
1. Bananas	81	30 lb.	$0.44 lb.	$ 13.20
2. Parsley	107	4 bunches	$0.80/bunch	$ 3.20
3. Oranges	101	3 cases	$31.50/case	$ 94.50
4. Lemons	35	6 cases	$29.20/case	$175.20
5. Cabbage	85	2 bags	$13.80/bag	$ 27.60
6.				
7.				
8.				
9.				
10.				
Total				$313.70

Order Date: 1/15	**Comments:** Please notify in advance if there are product outages
Ordered By: Joshua Jenn	
Received By: _____	**Order Transmitted by:** Joshua Jenn
Delivery Instructions: After 1:00 p.m. only	

Figure 8.6   Sample Purchase Order

From a cost control perspective, however, it is generally best to separate purchasing and receiving tasks in a way that one individual places purchase orders, and a different individual verifies delivery and acceptance of orders. Note: If this is not done, the potential for purchasing fraud or theft can be substantial.

The basic requirements for effective receiving are:

1) Proper location
2) Proper tools and equipment
3) Proper staff training

Proper Location

An operation's "back door," which is most often used for receiving, is frequently no more than a kitchen entrance. In fact, the receiving area must be adequate for receiving, or product loss and inconsistency may result.

The receiving area must be large enough to allow for properly checking products delivered against the **delivery invoice** (the seller's record of what is being delivered) and the PO (the buyer's record of what was originally ordered).

Key Term

Delivery invoice: A detailed listing of products and their costs delivered by a vendor.

In addition to space required to count and weigh incoming items, accessibility to equipment required to move products to proper storage areas and dispose of excess packaging is important. A location near refrigerated storage areas is desirable for maintaining deliveries of refrigerated and frozen products at their optimal temperatures.

The receiving area should be kept extremely clean to avoid contaminating incoming products. The area should be well lit and properly ventilated. Too little light may cause product defects to go unnoticed, and excessive heat in the receiving area can quickly damage delivered goods, especially if they are refrigerated or frozen products. Flooring in receiving areas should be light in color and of a type that is easily cleaned.

Proper Tools and Equipment

Although the tools and equipment needed for effective receiving vary by type and size of operation, some items are standard in any receiving operation, and they include:

- ✓ Scales: Scales should be of two types: those accurate to the fraction of a pound (for large items) and those accurate to the fraction of an ounce (for smaller items and pre-portioned meats).
- ✓ Wheeled equipment: These items, whether hand trucks or carts, should be available so goods can be moved quickly and efficiently to proper storage areas.
- ✓ Box cutters: These allow the receiving clerk to quickly remove excess packaging and accurately verify the quality of delivered products.
- ✓ Thermometers: Foods must be delivered at their proper storage temperatures. All operators should establish the range of temperatures they deem acceptable for product delivery. For most operators, these acceptable temperatures are shown in Figure 8.7.
- ✓ Calculator: A calculator should be available in case the original invoice is either increased or decreased in amount because of incorrect vendor pricing or because items on the invoice were not delivered. In addition, invoice totals change when all or a portion of the delivery is rejected because items were of substandard quality.

Item	°F	°C
Frozen Foods	10 °F or less	−12 °C or less
Refrigerated Foods	30 to 45 °F	−1 to 7 °C

Figure 8.7 Acceptable Temperatures for Delivered Products

Proper Staff Training

Receiving personnel should be properly trained so they can readily verify the key characteristics of delivered items:

1) Weight
2) Quantity
3) Quality
4) Price

Weight

Verifying product weight when delivered is not always easy. For example, a 14-pound package of ground beef will look exactly like a 15-pound package. There is no easy way to tell the difference without putting the product on a scale. Therefore, receiving clerks should be required to weigh all incoming meat, fish, and poultry. An exception to this rule can be unopened Cryovac (sealed) packages containing items such as hot dogs and bacon, and then the entire case should be weighed to detect content shortages.

Quantity

The proper counting of products is as important as proper weighing. Suppliers typically make more mistakes in *not* delivering products than they do in excessive deliveries. Products delivered but not charged for cost the supplier; products not delivered but charged for cost an operator money.

To illustrate the importance of count, assume an operator ordered five cases of gin. In this case, the operator will want to receive and pay for five cases. This is important for two reasons. First, the operator should only pay for products that have been delivered. Second, and just as important, if the operator has used sales forecasts to prepare their PO correctly, they truly need five cases of gin. If only three are delivered, they may not be able to make and serve enough gin-based drinks their guests will desire.

Quality

The quality of delivered items is just as important as the quantity of delivered items. If an operator takes time to develop product specs, but then accepts delivery of products that do not match the specifications, they are wasting their own time and effort, and risking the serving of lower-quality products to their guests. For these reasons, the product quality of delivered items should always be compared to the applicable product spec used by the operation.

Price

When the person responsible for purchasing food prepares a PO, a confirmed and quoted price should be carefully recorded. It is never safe to assume that the delivered price will match the price listed on the PO.

If the receiving clerk has access to the original PO, it is a simple matter to verify the quoted and delivered prices. If these numbers do not match, the operator should be notified immediately. If this is not possible, the driver and the receiving clerk should initial the Comments section of the PO, to indicate the difference in the two prices, and a **credit memo** should be prepared.

If the receiving clerk has no record of the quoted price from either a PO or an equivalent document, price verification of this type is not possible. An inability to verify the quoted delivered prices of products at time of delivery is most often an indication of a poorly designed receiving system.

Key Term

Credit memo: An adjustment to a vendor's delivery invoice that reconciles any differences between the delivery invoice, the products ordered (PO), and products received.

What Would You Do? 8.1

"Come on Sonya, give me a break. I'm already behind because my truck broke down this morning, and I've got people all over town calling my boss to scream about their deliveries. It wasn't my fault I'm late. I'm just the driver," said Mitchell.

Mitchell, who makes deliveries for Spartan Produce, was talking to Sonya, the kitchen manager and receiving clerk for the High-Five Restaurant. The delivery Mitchell was making was a big one, and it was two hours late. "You know, Sonya," Mitchell continued, "you folks take longer to accept a delivery than any other restaurant on my route. Nobody else inspects and weighs like you do. And you hardly ever find any problems. Look, I know it's a big delivery, but just this once can't you just sign the invoice and let me get going? I want to see my son's ball game, and I won't make it if you take forever to inspect this load."

"I don't know Mitch," replied Sonya. "We've got procedures to follow here, and I'm supposed to use them every time."

"Look, just sign the ticket. If you find a problem later, I'll take care of it. I promise you," said Mitchell, who appeared to be increasingly flustered.

Assume you were Sonya. How would you likely respond to Mitchell's request that you speed up? Now assume you were this restaurant's owner. If you were personally accepting the delivery, how would you likely respond to Mitchell's request?

Managing Inventory and Production

After receiving staff have properly accepted the food products an operation has purchased, the next step in product cost control is to properly store the items. In most cases, the inventory management process consists of two main tasks:

1) Placing products in storage
2) Maintaining product security

Placing Products in Storage

Many food products are highly perishable, and they must be moved quickly from the receiving to storage areas. This is especially true for refrigerated and frozen items. An item such as ice cream, for example, can deteriorate substantially in quality if it remains at room temperature for only a few minutes. To optimize product quality, in most cases a foodservice operation should utilize a **first-in first-out (FIFO) inventory system**.

When a FIFO inventory management system is in use, older products (first in) are used before newer products. Receiving clerks must also take great care to place new stock behind, underneath, or at the bottom of older stock.

FIFO is the preferred storage technique for most perishable and nonperishable items.

Key Term

First-in first-out (FIFO) inventory system: An inventory management system in which products already in storage are used before more recently delivered products.

Failure to implement a FIFO system of storage management can result in excessive product loss due to spoilage, shrinkage, or deterioration of product quality.

To maintain the quality and security of delivered products, they should be immediately placed into one of three storage areas:

✓ Dry-storage
✓ Refrigerated storage
✓ Frozen storage

Dry-storage

Dry-storage areas should be maintained at a temperature ranging between 65 °F and 75 °F (18 °C and 24 °C). Temperatures below that range can harm food products. More often, however, dry-storage temperatures can exceed those at the upper limits of temperature acceptability. This is often because storage areas are in poorly ventilated and closed-in areas of an operation, and excessively high or low temperatures will damage dry-storage products.

Shelving in dry-storage areas must be easily cleaned and sturdy to hold the weight of dry products. Slotted shelving is preferred over solid shelving for storing food because slotted shelving allows better air circulation around stored products.

All shelving should be placed at least 6 inches above the ground to allow for proper cleaning beneath the shelving and to ensure proper ventilation.

Refrigerated Storage

Time and temperature control for safety foods (TCS food) are those which must be carefully handled for time and temperature controls to keep them safe. Refrigerator temperatures used to store TCS foods should be maintained at 41 °F (5 °C) or less. In most cases, refrigerator thermostats can be set 40 °F (4.5 °C) to ensure proper cold food holding temperatures are maintained.

Key Term

Time and temperature control for safety foods (TCS foods): Foods that must be kept at a particular temperature to minimize the growth of food poisoning bacteria or to stop the formation of harmful toxins

Refrigerators should be opened and closed quickly when used to lower operating costs and to ensure that refrigerated items remain at their peak of quality. Refrigerators should be properly cleaned on a regular basis. Condensation drainage systems in refrigerators should be checked at least weekly to ensure they remain clean and function properly.

Freezer Storage

Freezer temperatures should be maintained at 0 °F (−18 °C) or less. Delivered products should be carefully checked with a thermometer when received to ensure they are solidly frozen and have been delivered at the proper temperature. Unless they are built in, frozen-food holding units and refrigerators should be high enough off the ground to allow for easy cleaning around and under them to prevent cockroaches and other insect pests from living beneath them. Stand-alone units should be placed 6–10 inches away from walls to allow for the free circulation of air around and efficient operation of the units.

Frozen-food holding units must be regularly maintained, a process that includes cleaning them inside and out, and constant temperature monitoring to detect possible improper operation. A thermometer permanently placed in the unit, or one easily read from outside the unit, is best. It is also helpful to periodically check that gaskets on freezer and refrigerator doors tightly seal the food cabinet. This helps reduce operating costs and maintains peak product quality for a longer period of time.

Find Out More

The United States Food and Drug Administration (FDA) produces and regularly updates a food code containing information about proper handling of foods stored and prepared for sale in foodservice operations.

> The 2017 edition of the food code made some significant changes to the 2013 edition, including a revised requirement that the Person in Charge (PIC) of a foodservice operation be a Certified Food Protection Manager (CFPM).
>
> To review this and other food safety recommendations included in the FDA food code, go to:
>
> www.fda.gov/Food/GuidanceRegulation/RetailFoodProtection/FoodCode

Maintaining Product Security

Items held in storage must be considered as if they were cash, and it will take cash to replace inventory items mishandled during storage. For example, if a stored item is stolen, that item's sales value in cash disappears as well. Nearly all foodservice establishments experience some amount of theft in its most strict sense and some employee theft is nearly impossible to detect. Even the most sophisticated, computerized control system cannot determine if, for example, an employee walked into the produce walk-in and ate several fresh grapes.

Similarly, an employee who takes home two small sugar packets per night will likely go undetected. In neither of these cases is the amount of loss significant, and certainly they do not create a need to install security cameras in the walk-in or to search employees as they leave at the end of their shift. What operators must do, however, is to make it difficult to remove significant amounts of food or beverage items from storage without authorization. Good cost control and **issuing systems** must be in place to achieve this goal.

Key Term

Issuing system: The procedures in place to control the removal of products from storage.

Issuing is the formal process of removing needed food and beverage products from inventory. In smaller properties, issuing may be as simple as entering a locked storeroom, selecting the needed product, and then re-locking the storeroom door when leaving. In a more complex operation, especially one that serves alcoholic beverages, this method would likely be inadequate to achieve appropriate control.

While the specific issuing system needs of a foodservice operation may differ, the removal of stored items should be based on management-approved estimates of future sales and production schedules. Maintaining effective product security can typically be achieved if the following key principles are observed:

✓ Food, beverages, and supplies are removed from storage only as needed and based on approved production schedules.

✓ Needed items should be issued only with management approval.

✓ If a written record of issues is kept, each person removing food, beverages, or supplies from the storage area must sign to acknowledge receipt of the products.
✓ Products that do not ultimately get used should be returned to the storage area, and their return should be recorded.

Managing Production

Effective management of the production process is essential for effective cost control. Generally, managing the production process in a foodservice operation means controlling four key areas:

✓ Waste
✓ Overcooking
✓ Over-portioning
✓ Improper carryover utilization

Waste

Product losses from food or beverage waste is one example of excessive food costs. Some waste may be easy to observe such as when an employee does not use a rubber spatula to remove all salad dressing from a 1-gallon jar. However, improper work of a salad preparation person can yield excessive amounts of chopped lettuce and a reduced amount of lettuce needed to prepare salads and higher per-portion salad costs.

Management must consistently demonstrate its concern for the value of products. Each employee should realize that wasting products directly affects the operation's profitability and, therefore, his or her own economic well-being.

In general, food waste results from poor training or an operator's inattentiveness to detail. Unfortunately, some operators and employees feel that small amounts of food waste are unimportant. Effective operators know that a primary goal in reducing production waste should be to maximize product utilization and minimize the dangerous "it's only a few pennies and so doesn't matter" syndrome.

Overcooking

Cooking is the process of exposing food to heat. Excessive cooking, however, most often results in reduced product volume regardless of whether the item cooked is roast beef or vegetable soup. The reason: Many foods have high moisture contents and heating usually results in moisture loss. To minimize this loss, cooking times and methods listed on standardized recipes must be carefully calculated and followed.

Excess heat is the enemy of both well-prepared foods and an operator's cost control efforts. Too much time on the steam table line or in the holding oven extracts moisture from products, reduces product quality, and yields fewer portions that are available for sale. The result is unnecessarily increased food costs.

To control product loss from overcooking, operators must strictly enforce standardized recipe cooking times. This is especially true for high moisture content items such as meats, soups, stews, and baked goods. Moreover, extended cooking times can result in total product loss if properly prepared items are placed in an oven, fryer, steam equipment, or broiler and then "forgotten" until it is too late! It is, therefore, advisable to provide kitchen production personnel with small, easily cleanable timers and thermometers for which they are responsible. These can help substantially in reducing product losses from overcooking.

Over-portioning

Perhaps no other area of food and beverage cost control has been analyzed and described as fully as control of portion size, and there are two reasons for this. First, over-portioning by service personnel increases operating costs and may lead to mismatches in production schedules and anticipated demand. For example, assume 100 guests are expected, and 100 products to be served to them are produced. However, over-portioning causes the operation to be "out" of the product after only 80 guests were served. The remaining 20 guests will not receive portions because over-portioning allowed their food to be served to other guests.

Secondly, over-portioning must be avoided because guests want to believe they receive fair value for what they spend. If portions are large one day and small the next, guests may believe they have been cheated on the second day. Consistency is a key to operational success in foodservice, and guests want to know exactly what they will receive for their money.

In many cases, inexpensive kitchen tools are available to help employees serve the proper portion size. Whether these tools consist of scales, food scoops, ladles, dishes, or spoons, employees require an adequate number of easily accessible portion control devices to use them consistently. The constant checking of portion sizes served is an essential management task. When incorrect portion sizes are noticed, they must be promptly corrected to avoid product cost increases.

Improper Carryover Utilization

As addressed earlier in this chapter, predicting guest counts is a necessary but inexact process. Most foodservice operators want to offer the same broad menu to the evening's last diner as was offered to its first guest, so it is inevitable that some prepared food will remain unsold at the end of the shift (called "carryovers," or "leftovers").

In some segments of the hospitality industry, carryovers are a potential problem area, but less so in other operations. Consider the operator of an upscale gelato shop. At the end of the day, any unsold gelato is simply held in the freezer until the next day with no measurable loss of either product quantity or quality. Contrast that situation, however, with a full-service cafeteria. If closing time is 8:00 p.m., management wishes to have a full product line, or at least some of each menu item, available to the guest who walks in the door at 7:55 p.m. Obviously, in five more minutes, many displayed items will become carryovers, and this cannot be avoided. A manager's ability to effectively integrate carryover items on subsequent days can make the difference between profits and losses in some operations.

In many cases, carryover foods cannot be sold for their original value. For example, today's beef stew made from yesterday's prime rib will not likely be sold at prime rib prices. Carryovers generally mean reduced income and less profits relative to product value, and it is critical to minimize carryovers.

Carryover items that can be re-used should be properly labeled, wrapped, and FIFO-rotated so the items can be found and re-used easily. This results in greater employee efficiency to locate carryover items, and reduced energy costs because refrigerator doors will be left open for shorter time periods . Most operators find that requiring foods to be properly labeled and stored in clear plastic containers helps to manage these procedures.

Controlling the Cost of Sales Percentage

In some cases, operators who evaluate their product usage determine that their operations are extremely effective at utilizing food and beverage products. In other cases, product-related costs are too high to allow them to meet their profit goals, and then corrective actions must be implemented.

To illustrate, assume an operator has a food cost percentage goal (target) of 35%. In one specific accounting period, however, the operator had a 38% cost of sales, and they must now reduce the excessive cost of sales percentage. Therefore, despite missing the target, this operator's goal should not simply be to reduce overall product cost. Rather, the goal should be to maintain current (or increasing!) sales levels while achieving product costs that represent their proper proportion of sales. This, in turn, requires foodservice operators to undertake three important tasks:

1) Controlling food costs
2) Controlling beverage costs
3) Optimizing cost of sales

Controlling Food Costs

Excessive food costs can result when control systems do not prevent losses incurred while products are in storage, losses due to overcooking or overportioning, and employee theft. It's especially important to have programs in place to minimize opportunities for product theft by employees. Note: Most

1) Keep all storage areas locked and secure.
2) Issue food only with proper authorization and management approval.
3) Monitor the use of all carryovers.
4) Do not allow food orders to be prepared unless a POS entry, guest check, or other written request precedes the preparation.
5) Maintain an active inventory control system.
6) Ensure all food received is signed for by the appropriate receiving employee.
7) Do not pay suppliers for food products without an appropriate and signed delivery invoice.
8) Do not allow employees to remove food from the site without management's specific approval.

Figure 8.8 Product Security Tips

kitchen-related theft involves removing food products (not money) from the premises, and in many operations employees have "easy" access to valuable food and alcoholic beverage products.

The product security tips shown in Figure 8.8 can be helpful when designing control systems to ensure the safety and security of food products.

Controlling Beverage Costs

Employee product theft can occur in either the bar or the kitchen production areas, but it is often more prevalent in bar areas. Indeed, bar theft is one of the most frequent types of thefts in the foodservice industry. Although it may be impossible to halt all kinds of bar theft, operators can help minimize opportunities for product theft in bar areas by watching for several bartender activities.

Order Filled but Not Rung Up

In this scenario, a bartender delivers a drink as requested by a guest or server, but the drink sale is never recorded in the POS system, and the bartender simply keeps the amount of the sale. As a result, the sales recorded in the POS system will be equal to the amount of money in the cash drawer, even though the beverage inventory has been reduced by the amount of product served in the drink.

All drink orders should be entered in the POS system *prior* to drink preparation to prevent this type of theft. Management's vigilance is critical to ensure no drink is prepared until *after* the order is properly recorded by entering it into the POS system.

Overpouring and Underpouring

Overpouring occurs when more alcohol than management desires is served in a drink. Underpouring occurs when less alcohol than management desires is served in a drink. When bartenders overpour, they are stealing from the operation. When they underpour, they are shortchanging the guest.

When they overpour, bar staff may be doing so for friends or the extra tips this activity may yield. When they underpour, they may be making up for drinks they have given away or sold but have not rung up. In either case, management must work to prevent these actions.

Product Dilution

Often called "watering down the drinks," product dilution theft involves adding water to a liquor to make up for amount of the spirit that has either been stolen or given away. It is especially easy to water down products such as gin, vodka, rum, or tequila because these clear spirits will not change color when water is added. Detection of this type of theft is often difficult so operators must be vigilant.

Product Substitution

When a specific brand of liquor has been ordered and paid for by a guest, it should be served to him or her. If, however, a bartender substitutes a less expensive liquor for the requested brand and charges the guest for higher priced liquor, the bartender may intend to keep the difference in the prices paid for the two items. This has the effect of shortchanging guests who paid higher prices for premium liquor drinks they did not receive.

Direct Theft

Alcohol is a highly desirable product for many employees; therefore, its direct theft is always a possibility. This is especially true in a secluded beverage service area in which bartenders have direct access to product inventory and ease of exit. Proper controls and strict rules limiting the access of employees to beer, wine, and liquor storage areas should help detect this type of theft.

Technology at Work

It is important for foodservice operators to take regular inventories to accurately calculate their cost of product used. Doing so manually can be extremely time-consuming, especially for those operations with extensive beverage offerings.

Bars and restaurants are increasingly using computerized bar inventory control systems to assist with taking their liquor inventories and to control their liquor cost. While there are a variety of bar inventory control systems on the market, most share the common goals of increasing the speed, ease, and efficiency of performing detailed liquor inventory.

Historically, operators have taken end of month inventory for accounting purposes to calculate their cost of goods sold. Today, increasing numbers of operators are using liquor inventory as a management tool to increase the quality and detail of information they use to manage their businesses. Taking accurate liquor inventory can be an effective control measure that can assist operators in reducing liquor inventory shrinkage (theft) and increasing sales.

To examine some offerings of companies that have developed software to assist in controlling beverage costs, enter "bar inventory control systems" in your favorite search engine and review the results.

Optimizing Cost of Sales

After an operator has determined the actual cost of sales (see Chapter 3), they may find that their food and beverage product costs, when expressed as a percentage of sales, are too high, and they must be reduced. Operators facing this situation can choose from a variety of solutions to this problem if they first consider the cost of sales equation in its algebraic form.

In its simplest terms, the cost of sales equation is:

$$A/B = C$$

where:

$A = $ Cost of Products Sold
$B = $ Sales
$C = $ Cost Percentage

Figure 8.9 shows how algebra affects the equation and what the rules communicate to foodservice operators.

Algebraic Rule	Foodservice Operator Takeaway
If A is unchanged and B increases, C decreases.	If product costs can be kept constant while sales increase, the product cost percentage goes down.
If A is unchanged and B decreases, C increases.	If product costs remain constant but sales decline, the cost percentage increases.
If A increases at the same proportional rate B increases, C remains unchanged.	If product costs go up at the same rate sales go up, the cost percentage will remain unchanged.
If A decreases and B is unchanged, C decreases.	If product costs can be reduced while sales remain constant, the cost percentage goes down.
If A increases and B is unchanged, C increases.	If product costs increase with no increase in sales, the cost percentage will go up.

Figure 8.9 Rules of Algebra and Foodservice Operator Takeaways

In general, foodservice operators must control the variables that impact the product cost percentage and reduce the overall value of "C" in the cost percentage equation. Basic product cost reduction approaches can be used to optimize their overall product cost percentage:

✓ Ensure all products purchased are sold
✓ Decrease portion size relative to price
✓ Vary recipe composition
✓ Alter product quality
✓ Achieve a more favorable sales mix
✓ Increase price relative to portion size

These strategies can be applied when there are excessive food costs and beverage costs. It is the careful selection and mixing of these approaches to cost control that differentiate the successful operator from his or her unsuccessful counterpart.

It is not the authors' contention that all product cost reduction methods are exhausted by these approaches. However, alternative approaches presented illustrate how analysis of cost percentage reduction alternatives can benefit all operators (not just beverage operators).

Ensure All Products Purchased Are Sold

This strategy has tremendous implications, and it includes all phases of professional purchasing, receiving, storage, inventory, issuing, production, service, and cash control. Perhaps the hospitality industry's greatest challenge in product cost control is ensuring that all products, once purchased, generate cash sales that are ultimately deposited into the operation's bank account.

Decrease Portion Size Relative to Price

Product cost percentages are directly affected by an item's portion cost (see Chapter 7), which is a direct result of portion size. Too often, foodservice managers and bar operators assume that their standard portion sizes must conform to some unwritten rule of uniformity. However, most guests would prefer a smaller portion size of higher quality ingredients than the reverse. In fact, one problem restaurants may have is that their portion sizes are too large. The result: excessive food loss because uneaten products left on plates must be thrown away. It is important to remember that portion sizes are solely determined by foodservice operators, and they are variable.

An example of the impact of portion size on product cost percentage is shown in Figure 8.10. This figure presents the significant effect on liquor (only) cost percentage of varying the standard drink size served in an operation using $21.00 per liter as the standard cost of liquor and assuming 0.8-ounce evaporation per 33.8-ounce (1-liter, or 1,000-ml) bottle, and a standard $10.00 selling price per drink.

Note in this example that the operator's liquor cost percentage ranges from 6.4%, when a 1-ounce portion is served, to 12.7%, when a 2-ounce portion is served.

Drink Size	Drinks per Liter	Cost per Liter	Cost per Drink	Sales per Liter	Liquor Cost % per Liter
2 oz.	16.5	$21.00	$1.27	$165.00	12.7%
1¾ oz.	18.9	$21.00	$1.11	$189.00	11.1%
1½ oz.	22.0	$21.00	$0.95	$220.00	9.5%
1¼ oz.	26.4	$21.00	$0.80	$264.00	8.0%
1 oz.	33.0	$21.00	$0.64	$330.00	6.4%

Figure 8.10 Impact of Drink Size on Liquor Cost Percentage at Constant Selling Price of $10.00 per Drink

Portion sizes of both food and drink items directly affect product cost percentages and guest perceptions of value delivered. As a result, when establishing portion sizes, operators should carefully consider all variables affecting their sales. These may include location, service levels, competition, and the clientele being served.

Vary Recipe Composition

Experienced foodservice operators know that the simplest standardized recipes can often be changed. For example, consider the proper ratio of clams to potatoes when preparing 100 servings of high-quality clam chowder. Since the cost of one pound of clams exceeds the cost of one pound of potatoes, 100 servings of clam chowder made with an increased amount of clams will cost more to produce than 100 servings of chowder made with increased amounts of potato. What constitutes an ideal recipe composition in this example? The answer must be addressed by the operator, and the answer to it and each standardized recipe used will impact the operations' overall food cost percentage.

Similarly, the proportion of alcohol to mixer affects beverage cost percentages. Sometimes the amount of alcohol in drinks can be reduced, and overall drink sizes can actually be increased. One example: Increasing the drink's proportion of lower cost standardized drink recipe ingredients such as milk, juices, and soda may enable a smaller portion of higher cost spirit products. Utilization of this beverage cost reduction strategy often contributes to a feeling of satisfaction by the guest, while allowing the operator to reduce beverage costs and increase profitability.

Alter Product Quality

In nearly all cases, higher-quality food and beverage products cost more than lower-quality products. Therefore, one way to reduce costs is to reduce product quality. This area must be approached with great caution because a foodservice operation should never serve products of unacceptable quality. Rather, cost-conscious operators should always purchase the quality of product appropriate for its intended use.

When operators determine that an appropriate ingredient, rather than the highest-cost ingredient, provides good quality and good value to guests, product costs might be reduced with product substitution. One caveat: Lower quality products may cost less, but customers may perceive that menu items made from these lower quality ingredients provide reduced levels of value.

Achieve a More Favorable Sales Mix

Experienced operators know that their customers' item selection decisions have a direct and significant impact on product cost percentages. The reason: An operation's overall product cost percentage is determined in large part by the operation's **sales mix**.

Sales mix affects overall product cost percentage anytime guests have a choice among several menu selections, with each item having a unique product cost percentage.

Key Term

Sales mix: The series of individual guest purchasing decisions that result in a specific overall food or beverage cost percentage.

To consider how sales mix directly affects an operation's overall product costs, assume only three menu items are being sold. In this operation, each item is priced separately, but the operation also offers a special "bundle meal" (see Chapter 7) that includes one of each item when purchased at the same time, as shown in Figure 8.11.

When reviewing Figure 8.11, it is easy to see that if, on a specific day, 100% of the operation's guests bought a hamburger and nothing else, the operation's product cost would be 37.6%. If, on another day, 100% of its customers purchased a soft drink and nothing else, the product cost percentage for that day would be 15.2%.

Similarly, if every guest purchased only the "Bundle Meal" on a specific day, the operation would achieve a 33.1% product cost. An operation's actual product cost is largely determined by the "mix" of the individual product costs resulting from the menu item choices made by the operation's guests.

Operators can directly influence guest selection and sales mix by techniques including strategic pricing, effective menu design, and creative marketing. However, the guests will always determine an operation's overall product cost percentage because they determine an operation's sales mix. Recognizing the impact

Menu Item	Product Cost	Selling Price	Product Cost %
Hamburger	$1.50	$3.99	37.6%
French fries	$0.50	$1.99	25.1%
Soft drink	$0.15	$0.99	15.2%
Bundle meal	$2.15	$6.49	33.1%

Figure 8.11 Four-item Menu

of sales mix can help operators better understand that effective marketing and promotion of lower-cost items can reduce their product cost percentage and increase profitability, while allowing portion size, recipe composition, and product quality to remain constant.

Increase Price Relative to Portion Size

Some operators facing rising product costs and the resulting cost percentages think the only tactic is to increase their menu's selling prices. While this can be done, this idea must be approached with great caution. There may be no bigger foodservice temptation than to raise prices to counteract management's ineffectiveness at controlling product costs!

There are times when selling prices must be increased. This is especially true in inflationary times or when there are unique product shortages. Price increases should be considered, however, only when all other alternatives and needed steps to control costs have been considered and effectively implemented.

Properly managing food and beverage product costs is complex and must be accomplished with the utmost skill. The goal of all purchasing, production, and product cost control systems should be to deliver to guests only high-quality products sold at a fair price. Doing this requires a team effort. Typically, the cost of labor to pay the team is second only to the cost of products in the operation of a foodservice facility. Note: In some operations, the cost of labor actually exceeds the cost of products. Controlling and accounting for labor cost is very important, and the ways to do so are the topic of the next chapter.

What Would You Do? 8.2

"I know you have been buying the 7-ounce bullion cup for your crab corn chowder, and I still have those. They cost $1.32 each, so the 36-cup case would cost you $47.52." said Dave Segoula, the sales representative for Commercial Kitchen Supplies.

David was talking to Sharon Pelley, the owner of the Seaside Shack, a casual restaurant located on the beach of the Mississippi gulf coast.

"O.K.," said Sharon, "that's what we use now. What are you suggesting?"

"Well," replied Dave, "we just started carrying a 6.5-ounce bullion cup. These are only $1.16 each or about $42 for the 36-cup case. You save money on the cups, but you'll really save money because you can reduce the portion size of your chowder, and no one will ever know!"

Assume you were Sharon. What would be the likely impact on your food cost percentage if you followed Dave's suggestion and bought the smaller 6.5-ounce cup? Would you agree with Dave that your guests would never notice that you are now using a smaller size cup? How would your answer to this key question affect your cup purchasing decision?

Key Terms

Sales forecast	Purchase order (PO)	Time and temperature
Popularity index	Delivery invoice	control for safety food
Product specification	Credit memo	(TCS food)
(spec)	First-in first-out (FIFO)	Issuing system
Count (product)	inventory system	Sales mix

Operator's 10-Point Tactics for Success Checklist

Evaluate your need for, and the current status of, each of the following operational tactics. For those tactics you think are important, but not yet in place, develop an action plan for its implementation including who will be responsible for the tactic's completion and the target date by which it should be completed.

				If Not Done	
Tactic	Don't Agree (Not Done)	Agree (Done)	Agree (Not Done)	Who Is Responsible?	Target Completion Date
1) Operator understands the importance to profitability of controlling product costs.	——	——	——		
2) Operator recognizes the importance of accurate sales forecasts in the effective control of product costs.	——	——	——		
3) Operator recognizes the importance of standardized recipes in the effective control of product costs.	——	——	——		
4) Operator comprehends why formal purchase orders (POs) are essential when placing orders with vendors and suppliers.	——	——	——		
5) Operator understands that the basic requirements for effective receiving are proper location, proper tools and equipment, and proper staff training.	——	——	——		

Tactic	Don't Agree (Not Done)	Agree (Done)	Agree (Not Done)	If Not Done	
				Who Is Responsible?	Target Completion Date
6) Operator has systems in place to properly maintain and secure dry, refrigerated, and frozen food storage areas.	——	——	——		
7) Operator has designed and implemented an effective issuing system appropriate for their own business.	——	——	——		
8) Operator understands the impact on product cost of waste, overcooking, over-portioning, and improper carryover utilization.	——	——	——		
9) Operator fully understands the control implications of the A/B = C product cost percentage formula.	——	——	——		
10) Operator recognizes the various actions and strategies that can reduce a product cost percentage to its targeted amount.	——	——	——		

9

Labor Cost Control

What You Will Learn
1) The Importance of Controlling Labor Costs
2) How to Record Labor Costs
3) How to Evaluate Labor Productivity

Operator's Brief

In this chapter, you will learn about controlling labor costs, an expense that, for most foodservice operations, is second in importance only to that of product costs. In fact, in some operations, labor costs exceed the amount spent on products.

The expenses of management, staff, and benefits comprise an operation's total labor costs. In addition to the cash amounts paid for these items, however, there are numerous non-cash factors that will affect an operation's total cost of labor. Examples include employee selection, training, supervision, scheduling, breaks, the menu, available equipment and tools, and desired levels of service. Each of these non-cash factors are examined in this chapter.

As you record and account for your labor costs, they can be classified several ways, including management and staff, fixed and variable payroll, and controllable and non-controllable labor expense. Special aspects of payroll accounting in the foodservice industry can include addressing issues of regular and overtime pay, child labor laws, and tip accounting.

To determine how much labor expense is needed to operate your business, productivity standards must be established. A variety of productivity measures to do so include sales per labor hour, labor dollars per guest served, and the labor cost percentage. Each of these measures are addressed in detail in the chapter including an examination of how they are calculated and their strengths and weaknesses. Finally, in this chapter, you will learn how to optimize your total cost of labor and how to use employee empowerment as a powerful labor cost reduction device.

The Importance of Labor Cost Controls

As addressed in the previous chapter, having the correct amount of food and beverage products available to serve guests is important. Having staff members needed to properly prepare and serve these products is also vital.

In years past, labor was relatively inexpensive. Today, however, in an increasingly costly labor market, foodservice operators must learn the scheduling and supervisory skills needed to maximize the effectiveness of their staff. They must also apply cost control skills to evaluate their efforts because labor is a significant expense in all foodservice operations and in some cases can exceed an operation's food and beverage costs.

In some sectors of the foodservice industry, a reputation for long hours, poor pay, and undesirable working conditions have caused some high-quality employees to look elsewhere for more satisfactory careers. It does not have to be that way. When labor costs are adequately controlled, foodservice operators will have the funds necessary to create desirable working conditions and pay wages attractive to the best employees. In every business, including foodservice, better employees mean better guest service and, ultimately, better profits.

Total Labor Costs

Prime costs (see Chapter 3) in a foodservice operation are the sum of its product (food and beverage) costs and its total labor costs. Therefore, when total labor expense is properly controlled or reduced, prime costs are also reduced.

When utilizing the Uniform System of Accounts for Restaurants (USAR), total labor costs included in an operation's prime cost calculations are the expenses for the operation's management, staff, and employee benefits, as illustrated in Figure 9.1.

Figure 9.1 Total Labor Costs

Typically, any employee who receives a salary is considered management. A **salaried employee** generally receives the same income per week or month regardless of the number of hours worked.

If a salaried employee is paid $1,500 per week when he or she works a complete week, that $1,500 is included in management expense. Salaried employees are more accurately described as **exempt employees** because their duties, responsibilities, and levels of decisions make them "exempt" from the overtime provisions of the federal government's Fair Labor Standards Act (FLSA).

Exempt employees do not receive overtime for hours worked in excess of more than 40 per week, and they are expected by most foodservice operators to work the number of hours needed to adequately perform their jobs.

Staff costs in a foodservice operation generally refers to the gross pay received by employees in exchange for their work. For example, if an employee earns $20.00 per hour and works 40 hours per week, the gross paycheck (the employee's paycheck before any mandatory or voluntary deductions) would be $800 ($20.00 per hour × 40 hours = $800). This gross amount is recorded as a staff cost.

Employee benefit costs (see Chapter 3) are incurred in every foodservice operation. Mandatory benefits such as required contributions to Social Security are included as an employee benefit cost, as are voluntary contributions such as an employer's contribution to an individual employee's 401K retirement account.

Key Term

Salaried employee: An employee who regularly receives a predetermined amount of compensation each pay period on a weekly, or less frequent, basis. The predetermined amount paid is not reduced because of variations in the quality or quantity of the employee's work.

Key Term

Exempt employee: Employees exempted from the provisions of the Fair Labor Standards Act. These workers typically must be paid a salary above a certain level and work in an administrative or professional position. The U.S. Department of Labor (DOL) regularly publishes a duties test that can help employers determine who meets this exemption.

The actual total amount of employment taxes and benefits paid by a specific operation can vary greatly. Expenses such as payroll taxes and contributions to workers' unemployment and workers' compensation programs are mandatory for all employers. Other benefit payments such as those for employee insurance and retirement programs are voluntary and vary based on the benefits a foodservice operation chooses to offer its employees. As employment taxes and benefit costs increase, an operation's total labor cost increases, even if its management and staff costs remain constant.

The difference between exempt and non-exempt employees is clear: Exempt employees are exempt from overtime pay. However, the definition of who qualifies as an exempt employee varies by state.

Employers must accurately classify employees as exempt or non-exempt. Misclassification can result in heavy fines and, without a firm grasp on the distinctions, an employer cannot accurately forecast future payroll costs.

Even experienced foodservice operators can sometimes make errors about the finer details of what makes an employee exempt or non-exempt. To gain a better understanding of who does and does not qualify as an exempt worker in a specific state, enter "exempt worker requirements (state name)" in your favorite search engine and review the results.

Factors Affecting Total Labor Costs

For some foodservice operators, analysis of total labor costs only involves an assessment of the salaries and wages they pay in cash, plus the cost of any benefits they provide. Experienced managerial accountants, however, recognize that total labor costs are directly affected by other important non-cash factors such as:

1) Employee selection
2) Training
3) Supervision
4) Scheduling
5) Breaks
6) Menu
7) Equipment/tools
8) Service level desired

Employee Selection

Choosing the right employee for an open position is vitally important in developing a highly productive workforce. Good foodservice operators know that proper employee selection procedures go a long way toward establishing the kind of workforce that is both efficient and cost-effective. This involves matching the right employee with the right job. The process begins with the development of the **job description**.

Key Term

Job description: A statement that outlines the specifics of a particular job or position within a company. It provides details about the responsibilities and conditions of the job.

A written job description should be maintained for every position in every foodservice operation. From the job description, an appropriate **job specification** can be prepared.

Job descriptions and job specifications are important because they enable foodservice operators to hire only employees who are qualified to do a job and do it well. Qualified workers can complete their tasks in a more cost-effective manner than will those who are not qualified.

Training

Perhaps no area under an operator's direct control holds greater promise for increased employee productivity than effective training. In too many cases, however, training in the hospitality industry is poor or almost non-existent. Highly productive employees are usually well-trained employees, and frequently employees with low productivity have been poorly trained. Every position in a foodservice operation should have a specific, well-developed, and ongoing training program.

Effective training improves job satisfaction and instills in employees a sense of well-being and accomplishment. It will also reduce confusion, product waste, and loss of guests. In addition, supervisors find that a well-trained workforce is easier to manage than one in which employees are poorly trained.

An operator's training programs need not be elaborate, but they must be consistent and continual. Foodservice employees can be trained in many areas. Skills training allows production employees to understand an operation's menu items and how they are best prepared. Service-related training may be undertaken, for example, to teach new employees how arriving guests should be greeted and properly seated.

It is important to recognize that training must be ongoing to be effective. Employees who are well trained in an operation's policies and procedures should be constantly reminded and updated if their skill and knowledge levels are to remain high. Performance levels can also decline because of a change in the operational systems or changes in equipment used. When these changes occur, employees must be retrained. Effective training costs a small amount of time in the short run but pays off extremely well in dollar savings in the long run.

Supervision

All employees require proper supervision, but this is not to say that all employees want to be told what to do. Proper supervision means assisting employees in improving productivity. In this sense, the supervisor is a coach and facilitator who provides employee assistance. Supervising should be a matter of assisting employees to do their best, not just identifying their shortcomings. It is said that employees think one of two things when they see their boss approaching:

1) Here comes help!

or

2) Here comes trouble!

For supervisors whose employees feel that the boss is an asset to their daily routine, productivity gains can be remarkable. Supervisors who only see their positions as one of power or who see themselves as taskmasters rarely maintain the quality workforce to compete in today's competitive marketplace.

It is important to remember that it is the employee, not management, who directly services guests. When supervision is geared toward helping employees, the guests and the entire operation benefits. When employees know that management is committed to providing high-quality customer service and will assist employees to deliver a high service level, worker productivity will improve.

Scheduling

Even with highly productive employees, poor employee scheduling by management can result in low productivity ratios. Consider the example in Figure 9.2, where an operator has determined two alternative schedules for pot washers for an operation that is open for three meals per day.

In Schedule A, four employees are scheduled for 32 hours at a rate of $15.00 per hour. Pot washer payroll, in this case, is $480 per day (32 hours/day × $15.00/hour = $480 per day). Each shift (breakfast, lunch, and dinner) has two employees scheduled.

In Schedule B, three employees are scheduled for 24 hours. At the same rate of $15.00 per hour, payroll is $360 per day (24 hours/day × $15.00/hour = $360 per day). Staff costs in this case are reduced by $120 ($480 − $360 = $120), and further savings will be realized because of reduced employment taxes, benefits, employee meal costs, and other labor-related expenses. Schedule A assumes that the amount of work to be done is identical at all times of the day. Schedule B covers both the lunch and the dinner shifts with two employees but assumes that one pot washer is sufficient in the early-morning period as well as very late in the day.

Scheduling efficiency during the day can often be improved with a **split-shift**, a technique used to match individual employee work shifts with peaks and valleys of customer demand. When utilizing split-shifts, an operator would, for example, require an employee to work a busy lunch period, be off in the afternoon, and then return to work for the busy dinner period.

Key Term

Split-shift: A working schedule comprising two or more separate periods of duty in the same day.

Schedule A

	7:30 to 8:30	8:30 to 9:30	9:30 to 10:30	10:30 to 11:30	11:30 to 12:30	12:30 to 1:30	1:30 to 2:30	2:30 to 3:30	3:30 to 4:30	4:30 to 5:30	5:30 to 6:30	6:30 to 7:30	7:30 to 8:30	8:30 to 9:30	9:30 to 10:30	10:30 to 11:30
Employee 1																
Employee 2																
Employee 3																
Employee 4																

Total Hours = 32

Schedule B

	7:30 to 8:30	8:30 to 9:30	9:30 to 10:30	10:30 to 11:30	11:30 to 12:30	12:30 to 1:30	1:30 to 2:30	2:30 to 3:30	3:30 to 4:30	4:30 to 5:30	5:30 to 6:30	6:30 to 7:30	7:30 to 8:30	8:30 to 9:30	9:30 to 10:30	10:30 to 11:30
Employee 1																
Employee 2																
Employee 3																

Total Hours = 24

Figure 9.2 Two Alternative Schedules

Increasingly, foodservice operators can utilize cloud-based scheduling systems designed to allow them to place the right number of workers in the right shifts on the right days. Most of these systems also include features that, with management's pre-approval, allow employees to pick up, drop, or swap shifts with other workers. These adaptions give staff more flexibility about when and how much they will work. This type of work flexibility is highly valued by employees and can reduce turnover rates caused by inconvenient worker scheduling and reduce staffing costs.

Breaks

Most employees cannot work at top speed for eight consecutive hours, and they have both a physical and mental need for work breaks. Frequent short breaks allow them to pause, collect their thoughts, converse with their fellow employees, and prepare for the next work session. Employees given these short breaks will outproduce those who are not given any breaks.

Federal law does not mandate that all employees be given breaks, but some states do. As a result, foodservice operators often must determine both the best frequency and length of designated breaks. In some cases, and especially regarding the employment of students and minors, both federal and state laws may mandate special workplace break requirements. Professional operators must be familiar with details about these laws if they apply to their businesses.

Menu

A major factor in employee productivity is an operation's actual menu. The menu items to be served have a significant effect on employees' ability to produce the items quickly and efficiently.

In most cases, the greater the number of menu items a kitchen must produce, the less efficient that kitchen will be. Of course, if management does not provide guests with enough choices, loss of sales may result. Clearly, neither too many nor too few menu choices should be offered. The question for operators most often is, "How many selections are too many?" The answer depends on the operation, its employees' skill levels, and the variety of menu items operators believe is necessary to properly serve their guests.

It is extremely important that the menu items selected by management are those items that can be prepared efficiently and serviced well. If done, worker productivity rates will be high, and so will guest satisfaction.

Equipment/Tools

Foodservice productivity ratios have not increased as much in recent years as have those of other businesses. Much of this is because foodservice is a labor-intensive rather than machine-intensive industry.

In some cases, equipment improvements have made kitchen work easier. Slicers, choppers, and mixers have replaced human labor with mechanical labor. However, in most cases robotics and automation do not significantly contribute to the foodservice industry.

Nonetheless, it is critical for operators to understand the importance of a properly equipped workplace and how it improves productivity. This can be as simple as understanding that a sharp knife cuts more safely, quickly, and better than a dull one or as complex as deciding which Internet system is best to provide data and communication links to the 1,000 stores in a quick-service restaurant chain. In all cases, it is an important part of every operator's job to provide employees with the tools needed to do their jobs quickly and effectively.

Service Level Desired

The average quick-service restaurant (QSR) employee normally serves more guests in an hour than the fastest server at an exclusive fine dining restaurant. The reason: QSR guests desire speed, not extended levels of rendered services. In contrast, fine dining guests expect more elegant and personal service that is delivered at a much higher level, and this increases the number of necessary employees.

After operators fully recognize the non-cash payment factors that impact their total labor costs, they must next understand the different types of payment-related labor expenses they will incur. Operators must also evaluate the productivity of their workers to determine if their labor dollars were well spent.

Accounting for Total Labor Costs

The major components of a foodservice operation's total labor cost can be viewed and accounted for in several ways. Payroll (see Chapter 3) is one major component of every foodservice operation's total labor costs. Some employees are needed simply to open the doors for minimally anticipated business. For example, in a small operation, payroll may include only one supervisor, one server, and one cook. The cost of providing payroll to these three individuals would be the operation's minimum payroll.

Assume, however, that this operation anticipated much greater than minimum business volumes. The increased number of expected guests means that the operation will need more cooks and servers, as well as cashiers, dish room personnel, and, perhaps, more supervisors to handle the additional workload. These additional positions create a work group that is far larger than the minimum staff, but it is needed to adequately serve the anticipated number of guests. In this scenario, payroll costs will increase.

Types of Labor Costs

Payroll costs may be viewed in several ways, including being either fixed or variable. **Fixed payroll** most often refers to the amount an operation pays in salaries. This amount is typically fixed because it remains unchanged from one pay period to the next unless the salaried employee separates employment from the organization or is given a raise.

Variable payroll consists primarily of those dollars paid to hourly employees (staff) and is the amount that should vary with changes in sales volume. Generally, as sales volume increases, variable payroll expense increase. In many cases, foodservice operators have little control over their fixed payrolls, but they have nearly 100% control over variable payroll expenses above their minimum staff levels.

Payroll costs may also be viewed as being either controllable or non-controllable. Payroll expenses that are wholly or partially controlled by management are considered **controllable labor expenses**, and labor costs beyond the direct control of management are **non-controllable labor expenses**. Examples of non-controllable labor expenses include employment taxes and some mandatory benefits.

Key Term

Fixed payroll: The amount an operation pays for its salaried workers.

Key Term

Variable payroll: The amount an operation pays for those workers compensated based on the number of hours worked.

Key Term

Controllable labor expenses: Those labor expenses under the direct influence of management.

Key Term

Non-controllable labor expenses: Those labor expenses not typically under the direct influence of management.

Accounting for Labor Costs

Properly accounting for labor costs is important because a variety of laws regulate how and when employees must be paid. For many foodservice operators, particular attention must be paid to proper accounting in two key payment-related areas:

✓ Regular and Overtime Pay
✓ Accounting for Tips

Regular and Overtime Pay

In the United States, the Fair Labor Standards Act (FLSA) establishes minimum wage, overtime pay, recordkeeping, and youth employment standards affecting employees in all businesses employing two or more workers. Covered non-exempt workers are entitled to be paid no less than the federal minimum per hour. Overtime pay at a rate not less than one and one-half times the regular rate of pay is required after 40 hours of work in a workweek. The FLSA sets worker pay requirements in several important areas:

Minimum wage

The FLSA establishes a minimum hourly rate that must be paid to all employees. If a foodservice operation is in a state with minimum wage higher than the federal minimum wage, the state minimum wage must be paid.

Overtime

Covered non-exempt employees must receive overtime pay for hours worked over 40 per workweek (any fixed and regularly recurring period of 168 hours—seven consecutive 24-hour periods) at a rate not less than one and one-half times the regular rate of pay.

There is no federal limit on the number of hours employees 16 years or older may work in any workweek. The FLSA does not require overtime pay for work on weekends, holidays, or regular days of rest unless overtime is worked on such days.

Hours Worked

Hours worked ordinarily include all the time during which an employee is required to be on the employer's premises, on duty, or at a prescribed workplace.

Recordkeeping

Employers must display an official poster outlining the requirements of the FLSA. Employers must also keep accurate employee time and pay records.

Child Labor

Child labor provisions set by the FLSA are designed to protect the educational opportunities of minors and prohibit their employment in jobs and under conditions detrimental to their health or well-being.

Every state has its own laws specifically dealing with child labor issues. When federal and state standards are different, the rules that provide the most protection to youth workers apply. Employers must comply with both federal law and applicable state laws.

Find Out More

Many foodservice operators employ young workers. The FLSA generally sets 14 years old as the minimum age for employment, and it limits the number of hours worked by minors under the age of 16. Special rules also apply to the specific types of tasks these workers can be assigned. Child labor laws also vary by state.

You can better understand the specific requirements of employing young people in a business by making an Internet visit to the department or agency responsible for enforcing youth employment laws in your state. To do so, enter "child labor laws in (state name)" in your favorite search engine, and then view the results to learn about any special laws or regulations that apply to minors working in the state.

Accounting for Tips

Foodservice operations are somewhat unique in that they are allowed to consider employee tips as part of their wage payments when satisfying FLSA worker pay requirements. As a result, a challenge faced by foodservice operators relates to the accounting procedures required to compensate their tipped employees fairly and legally and to keep records of doing so.

In most foodservice operations, a modern point-of-sale (POS) system is used to record the credit card and cash tips guests intend to give to their servers. Increasingly, however, some foodservice operations use a **tip distribution program** to manage the compensation of their tipped employees.

For example, when a tip distribution program is in place, if a customer gives the server a $50.00 tip on a $200 guest check that included $150.00 of food and $50.00 of alcoholic beverages, the program will assign a portion of the tip to the customer's server and another portion to the bartender(s) who prepared the customer's drinks.

Tip distribution programs are typically needed when an operation has either a **tip sharing** or **tip pooling** system in place.

The best tip distribution programs are interfaced directly to a restaurant's payroll accounting system because, if an employee "customarily and regularly" makes more than $30 per month in tips, then under current federal law that employee is considered a tipped employee for minimum wage and overtime pay purposes.

Key Term

Tip distribution program: A system of tip payment that allows an operation to distribute a customer's tip from an employee who actually received it to others who also provided service to the customer.

Key Term

Tip sharing: A tip system that takes the tips given to one group of employees and gives a portion of them to another group of employees: used, for example, when server tips are shared with those who bus the server's tables.

Key Term

Tip pooling: A tip system that takes the tips given to individual employees in a group and shares them equally with all other members of the group: used, for example, when bartender tips given to an individual bartender are shared equally with all bartenders working the same shift.

Many states also have special laws for tipped employees, and some have different standards for qualification as a tipped employee than the federal standard.

Technology at Work

Tip distribution software, also referred to as tip pooling or gratuity management software, automates the process of paying tips to tipped workers at the end of their shifts and reduces the need for physical cash payouts.

These tools eliminate hours of manual labor by tracking the number of hours worked and automatically distributing tips to workers' bank accounts. These tools reduce payroll burden and instances of tip disparity due to an error or theft and expedite paying out tips to employees by connecting securely to their bank accounts or debit cards.

Additionally, most tip distribution software allows operators to manage tips across multiple locations and utilize proper reporting features to mitigate risks of non-compliance in tip payments.

To examine some of the offerings of companies that have developed software to assist in tip payment and proper recording, enter "tip distribution software" in your favorite search engine and review the results.

What Would You Do? 9.1

Ishaan Patel is the kitchen manager at the Tapron Corporation International Headquarters. The facility he helps manage serves 3,000 employees per day. Ishaan very much needs an additional dishwasher right away. Despite advertising his vacant position for several weeks, Ishaan has had few applicants. He is now interviewing Daniel, who is an excellent candidate with five years of experience and who is now washing dishes at the nearby Downriver Rustic Steakhouse.

Ishaan normally starts new dishwashers at $14.00 per hour. Daniel states that he currently makes $16.25 per hour, a rate that is higher than all but one of Ishaan's current dishwashers, many of whom have as much experience as Daniel.

As they end the interview, Daniel says that while he would very much like to work at Tapron, he simply will not leave his current job to take a "pay cut."

Assume you were Ishaan. Would you hire Daniel at a pay rate higher than most of your current employees? If so, what would you say to your current dishwashing employees if in the future Daniel shared his pay information with them? If not, what would be your plan for filling your vacant dishwasher position?

Assessment of Total Labor Costs

To determine how much labor expense is needed to properly operate their businesses, managers must be able to determine how much work is to be done and how much of it each employee can accomplish. If too few employees are scheduled to work, poor service and reduced sales can result because guests may choose to go elsewhere in search of superior service levels. If too many employees are scheduled, staff wages and employee benefits costs will be too high with the result of reduced profits.

To properly determine the number of workers needed, operators must have a good understanding of the **productivity** of each of their employees.

There are several ways to assess labor productivity. In general, productivity is measured by calculating a ratio (see Chapter 4) of productivity as shown below:

Key Term

Productivity (worker): The amount of work performed by an employee within a fixed time period.

$$\frac{\text{Output}}{\text{Input}} = \text{Productivity Ratio}$$

To illustrate use of this ratio, assume a foodservice operation employs 4 servers, and it serves 80 guests. Using the productivity ratio formula, the output is guests served, and the input is servers employed:

$$\frac{80 \text{ guests}}{4 \text{ servers}} = 20 \text{ guests per server}$$

This formula states that, for each server employed, 20 guests can be served. The productivity ratio is 20 guests to 1 server (20 to 1) or, stated another way, 1 server per 20 guests (1/20).

There are several ways of defining foodservice output and input, and there are several types of productivity ratios.

Productivity ratios can help an operator determine the answer to the key question, "How much should I spend on labor?" The answer to the question becomes more complex, however, when it is recognized that back-of-house (areas not open to public access) and front-of-house (areas within a food operation that are open to public access) productivity levels should, in most cases, be measured differently.

While each foodservice operation is unique, typical measures used to assess *back-of-house* productivity can include:

✓ Number of covers (guest orders) completed per labor hour
✓ Number of guest checks (table orders) processed per labor hour
✓ Average guest check completion time (in minutes)
✓ Number of menu items produced per hour worked
✓ Number of improperly cooked items (mistakes) produced per hour worked

Examples of typical measures used to assess the *front-of-house* productivity of servers include:

✓ Number of guests (not tables) served per server hour worked
✓ Number of menu "specials" sold per server hour worked
✓ Number of errors (voided sales) produced per shift worked
✓ Average guest check size (per guest served)

Regardless of the productivity measures used, foodservice operators must develop their own methods for managing payroll costs because every foodservice unit is different. Consider, for example, the differences between managing payroll costs incurred by a food truck operator and those required for a large banquet kitchen located in a 1,000-room convention hotel.

Although methods used to manage payroll costs may vary, payroll costs can and must be managed. While there are several ratios operators can use to assess their labor costs worker productivity, three commonly utilized are:

1) Sales per labor hour
2) Labor dollars per guest served
3) Labor cost percentage

Sales per Labor Hour

It is said that the most perishable commodity any foodservice operator can buy is the labor hour. When labor is not productively used, it disappears forever. It cannot be "carried over" to the next day as can an unsold head of lettuce or a slice of turkey breast. For this reason, some foodservice operators measure labor productivity in terms of the amount of sales generated for each labor hour used. This productivity measure is referred to as **sales per labor hour**.

The formula used to calculate sales per labor hour is:

Key Term

Sales per labor hour: The dollar value of sales generated for each labor hour used.

$$\frac{\text{Total Sales Generated}}{\text{Labor Hours Used}} = \text{Sales per Labor Hour}$$

When using this productivity measure, labor hours used is simply the sum of all labor hours paid for by an operation in a specific sales period. To illustrate, consider the operator whose four-week labor usage and the resulting sales per labor hour information is presented in Figure 9.3.

In this example, sales per labor hour ranged from a low of $19.50 in week 1 to a high of $28.66 in week 4. Sales per labor hour varies with changes in selling prices,

Week	Total Sales	Labor Hours Used	Sales per Labor Hour
1	$18,400	943.5	$19.50
2	21,500	1,006.3	21.37
3	19,100	907.3	21.05
4	24,800	865.3	28.66
Total	$83,800	3,722.4	$22.51

Figure 9.3 Four-Week Sales per Labor Hour

but it will not vary based on changes in the prices paid for labor. In other words, increases and decreases in the price paid per hour of labor will not affect this productivity measure. As a result, a foodservice operation paying its employees an average of $15.00 per hour could, using this type of measure for labor productivity, have the same sales per labor hour as a similar unit paying $20.00 for each hour of labor used. Obviously, the operator paying $15.00 per hour has paid far less for an equally productive workforce if the sales per labor hour used are identical in the two units.

Many operators like utilizing the sales per labor hour productivity measure because records on both the numerator (total sales) and the denominator (labor hours used) are readily available. However, depending on the record-keeping system employed, it may be more difficult to determine total labor hours used than total labor dollars spent. This is especially true when large numbers of managers or supervisors are paid by salary rather than by the hour. Note: In most cases, the efforts of both salaried workers and hourly paid staff should be considered when computing an operation's overall sales per labor hour.

Labor Dollars per Guest Served

Some foodservice operators measure labor productivity in terms of the labor dollars spent for each guest served. This productivity measure is referred to as **labor dollars per guest served**.

The formula used to calculate labor dollars per guest served is:

Key Term

Labor dollars per guest served: The dollar amount of labor expense incurred to serve each of an operation's guests.

$$\frac{\text{Total Cost of Labor}}{\text{Total Number of Guests Served}} = \text{Labor Dollars per Guest Served}$$

To illustrate, consider the operator whose four-week labor cost and the resulting sales per labor hour information is presented in Figure 9.4.

Week	Cost of Labor	Guests Served	Labor Dollars per Guest Served
1	$7,100	920	$7.72
2	8,050	1,075	7.49
3	7,258	955	7.60
4	6,922	1,240	5.58
Total	$29,330	4,190	$7.00

Figure 9.4 Four-Week Labor Dollars per Guest Served

In this example, the labor dollars expended per guest served for the four-week period would be computed as:

$$\frac{\$29,330}{\$4,190} = \$7.00$$

Note that, in this example (for three weeks, i.e., weeks 1–3), the operator provided guests with more than $7.00 of guest-related labor costs per guest served, but in the fourth week that amount fell to less than $6.00 per guest. This productivity measure, when averaged, can be useful for operators who find that their labor dollars expended per guest served are lower when their volume is high and higher when their volume is low.

The utility of labor dollars per guest served is limited in that it will vary based on the price paid for labor. Unlike sales per labor hour, however, it is not affected by changes in menu prices.

Labor Cost Percentage

Perhaps the most commonly used measure of employee productivity in the food service industry is the **labor cost percentage**.

The formula used to calculate labor cost percentage is:

$$\frac{\text{Total Cost of Labor}}{\text{Total Revenue Generated}} = \text{Labor Cost}\%$$

Key Term

Labor cost percentage: A ratio of overall labor costs incurred relative to total revenue generated.

A labor cost percentage allows an operator to measure the relative cost of labor used to generate a known quantity of sales. It is important to realize, however, that different operators may choose different methods of calculating this popular productivity measure.

Since a foodservice operation's total labor cost consists of management, staff, and employee benefits costs, some operators may calculate their labor cost percentage using only hourly staff wages, or staff wages and salary costs, but not benefit costs. This approach makes sense if an operator can directly control employee pay but not employee benefit costs. It is important to recognize, however, that if an operator wishes to directly compare their own labor cost percentage to that of other operations, both must have utilized the same formula.

Controlling the labor cost percentage is extremely important in the foodservice industry because it is often used to assess the effectiveness of management. If an operation's labor cost percentage increases beyond what is expected, management will likely be held accountable by the operation's ownership.

Labor cost percentage is a popular measure of productivity, in part, because it is so easy to compute and analyze. To illustrate, consider the case of Raelynn, a foodservice manager in charge of a casual service restaurant in a year-round theme park. The unit is popular and has a $20 per guest check average. Raelynn uses only payroll (staff wages and management salaries) when determining her overall labor cost percentage because she does not have easy access to the actual amount of taxes and benefits provided to her employees. Raelynn's own supervisor considers these labor-related expenses to be non-controllable and, therefore, beyond Raelynn's immediate influence.

Raelynn has computed her labor cost percentage for each of the last four weeks using her modified labor cost percentage formula. Her supervisor has given Raelynn a goal of 35% labor costs for the four-week period. Raelynn feels that she has done well in meeting that goal. Figure 9.5 shows Raelynn's four-week performance.

Using her labor cost percentage formula and the data in Figure 9.5, Raelynn's labor cost is calculated as:

$$\frac{\text{Cost of Labor}}{\text{Total Sales}} = \text{Labor Cost}\%$$

Or

$$\frac{\$29,330}{\$83,800} = 35\%$$

While Raelynn did achieve a 35% labor cost for the four-week period, Mandy, her supervisor, is concerned because she received several negative comments in

Week	Cost of Labor	Total Sales	Labor Cost %
1	$7,100	$18,400	38.6%
2	8,050	21,500	37.4
3	7,258	19,100	38.0
4	6,922	24,800	27.9
Total	**$29,330**	**$83,800**	**35.0**

Figure 9.5 Raelynn's Four-Week Labor Cost Percentage Report

week 4 regarding poor service levels in Raelynn's unit. Some of these were even posted online, and Mandy is concerned about the postings' potential impact on future visitors to the park's foodservice operations. When she analyzes the numbers in Figure 9.5, she sees that Raelynn exceeded her goal of a 35% labor cost in weeks 1 through 3 and then reduced her labor cost to 27.9% in week 4.

Although the monthly overall average of 35% is within budget, Mandy knows all is not well in this unit. Raelynn elected to reduce her payroll in week 4, but the negative guest comments suggest that reduced guest service resulted from too few employees on staff to provide the necessary guest attention. As Mandy recognized, one disadvantage of using an overall labor cost percentage is that it can hide daily or weekly highs and lows.

In Raelynn's operation, labor costs were too high the first three weeks, and too low in the last week, but she still achieved her overall target of 35%. Raelynn's labor cost of 35% indicates that, for each dollar of sales generated, 35 cents was paid to the employees who assisted in generating those sales. In many cases, a targeted labor cost percentage is viewed as a measure of employee productivity and, to some degree, management's skill in controlling labor costs.

While it is popular, in addition to its tendency to mask productivity highs and lows, the labor cost percentage has some additional limitations as a measure of productivity. Note, for example, what happens to this measure of productivity if all of Raelynn's employees are given a 5% raise in pay. If this were the case, her labor cost percentages for last month would be calculated as shown in Figure 9.6.

Note that labor now accounts for 36.8% of each sales dollar, but one should realize that Raelynn's workforce did not become less productive simply because they got a 5% increase in pay. Rather, the labor cost percentage changed due to a change in the price paid for labor. When the price paid for labor increases, labor cost percentage increases and, similarly, when the price paid for labor decreases, the labor

Week	Original Cost of Labor	5% Pay Increase	Cost of Labor (with 5% pay increase)	Total Sales	Labor Cost %
1	$7,100	$355.00	$7,455.00	$18,400	40.5%
2	8,050	402.50	8,452.50	21,500	39.3
3	7,258	362.90	7,620.90	19,100	39.9
4	6,922	346.10	7,268.10	24,800	29.3
Total	29,330	1,466.50	30,796.50	83,800	36.8

Figure 9.6 Raelynn's Four-Week Revised Labor Cost Percentage Report (Includes 5% Pay Increase)

Week	Cost of Labor	Original Sales	5% Selling Price Increase	Sales (with 5% selling price increase)	Labor Cost %
1	$7,100	$18,400	$920	$19,320	36.7%
2	8,050	21,500	1,075	22,575	35.7
3	7,258	19,100	955	20,055	36.2
4	6,922	24,800	1,240	26,040	26.6
Total	29,330	83,800	4,190	87,990	33.3

Figure 9.7 Raelynn's Four-Week Revised Labor Cost Percentage Report (Includes 5% Increase in Selling Price)

cost percentage decreases. Therefore, using the labor cost percentage alone to evaluate workforce productivity can sometimes be misleading.

Another example of the limitations of the labor cost percentage as a measure of labor productivity can be seen when selling prices are increased. Return to the data in Figure 9.5 and assume that Raelynn's unit raised all menu prices by 5% effective at the beginning of the month. Figure 9.7 shows how this increase in her selling prices would affect her labor cost percentage.

Note that increases in selling prices (assuming no decline in guest count or changes in guests' buying behavior) will result in decreases in the labor cost percentage. Alternatively, lowering selling prices without increasing total revenue by an equal amount will result in an increased labor cost percentage.

Although labor cost percentage is easy to compute and widely used, it is difficult to use as a measure of productivity over time. The reason: it depends on labor dollars spent and sales dollars received for its computation. Even in relatively noninflationary times, wages do increase, and menu prices are adjusted. Both activities directly affect labor cost percentage, but not worker productivity. In addition, institutional foodservice settings, which often have no daily dollar sales figures to report, can find that it is not easy to measure labor productivity using labor cost percentage because operators generally calculate and report guest counts or number of meals served rather than sales dollars earned.

Figure 9.8 summarizes key characteristics of the three measures of labor productivity presented.

Regardless of the productivity measure utilized, if an operator finds labor costs are too high relative to sales produced, problem areas must be identified, and corrective action must be taken. If the overall productivity of employees cannot be improved, other action(s) become important.

Measurement	Advantages	Disadvantages
Sales per Labor Hour = $\dfrac{\text{Total Sales}}{\text{Labor Hours Used}}$	1) Fairly easy to compute 2) Does not vary with changes in the price of labor	1) Ignores price per hour paid for labor 2) Varies with changes in menu selling price
Labor Dollars per Guest Served = $\dfrac{\text{Cost of Labor}}{\text{Guests Served}}$	1) Fairly easy to compute 2) Does not vary with changes in menu selling price 3) Can be used by non-revenue-generating units	1) Ignores average sales per guest and, therefore, total sales 2) Varies with changes in the price of labor
Labor Cost% = $\dfrac{\text{Cost of Labor}}{\text{Total Sales}}$	1) Easy to compute 2) Most widely used	1) Hides highs and lows 2) Varies with changes in price of labor 3) Varies with changes in menu selling price

Figure 9.8 Productivity Measures Summary

The approaches operators can take to reduce labor-related costs are different for fixed payroll costs than for variable payroll costs. Figure 9.9 summarizes strategies operators can use to reduce labor-related expense percentages in each of these two categories. Note that operators can only decrease variable payroll expense by increasing productivity, improving the scheduling process, eliminating employees, or reducing wages paid.

Labor Category	Actions
Fixed	1) Increase sales volume. 2) Combine jobs to eliminate fixed positions. 3) Reduce wages paid to fixed-payroll employees.
Variable	1) Improve productivity. 2) Schedule appropriately to adjust to changes in sales volume. 3) Combine jobs to eliminate variable positions. 4) Reduce wages paid to variable employees.

Figure 9.9 Reducing Labor-Related Expense

Another tactic often ignored can also increase employee productivity and reduce labor-related expense: **employee empowerment**. Employee empowerment results from a decision by management to fully involve employees in the decision-making process as far as guests and the employees themselves are concerned. Many experienced managers remember that it was once customary for management to (a) make all decisions regarding every facet of the operational aspects of its organization and (b) present them to employees as inescapable facts to be accomplished. Instead, an alternative approach occurs when employees are given the "power" to get involved.

Employees can be empowered to make critical decisions concerning themselves and, most importantly, the guests. Many foodservice employees work closely with guests, and numerous problems are more easily solved when employees are given the power to make it "right" for the guests. Successful operators often find that well-planned and consistently delivered training programs can be helpful. Empowered employees can also yield a loyal and committed workforce that is more productive, is supportive of management, and will "go the extra mile" for guests. Doing so helps reduce labor-related costs, builds repeat sales, and increases profits.

The previous chapter, as this chapter, focused on controlling and optimizing product and labor costs. Doing so allows foodservice operators to better understand what these costs *are*. The next chapter addresses the development of operating budgets; the process operators use to determine what these and other important costs *should be*.

Technology at Work

Properly accounting for payroll costs in a foodservice operation is a complex process. Foodservice operators with payroll and related accounting responsibilities recognize the impacts of numerous laws. These include those related to child labor, state minimum wages, overtime pay provisions, tip allocations, and applicable employment tax withholdings.

Fortunately, foodservice operators can obtain help with these important tasks by using payroll accounting software programs developed specifically for restaurants. The best of these programs interface directly with an operation's point-of-sale (POS) system. This permits easy calculation and payment of tips and the use of historical sales records and sales forecasts to help operators as they create future work schedules for employees.

To review some features included in these important management tools, enter "payroll accounting software for restaurants" in your favorite search engine and view the results.

What Would You Do? 9.2

"I'm just saying that we are already reporting our daily labor cost percentage to the owners, and now they also want to know this every day. It just seems like more paperwork to me," said Frank, the assistant manager of the Philadelphia Cheesesteak restaurant.

Frank was talking to Loretta, the restaurant's manager. Loretta had just informed Frank that the restaurant's owners wanted her to submit a daily "Average Drive-Through Ticket Time Report" along with the day's revenue and labor cost percentage reports.

Along with on-site dining, the Philadelphia Cheesesteak's drive-through was seeing an increasing amount of business. However, there have been some complaints about slow service. In response, the restaurant's owners requested that Loretta calculate an average drive-through ticket time. Note: Ticket time begins when a guest's order is placed, and it ends when the guest leaves the drive-through window with their ordered items.

This restaurant's POS system creates a unique guest check for each order, and the system records the time at which a guest's order was placed and when the order was completed. One reporting feature of the POS system is that it automatically calculates the average drive-through ticket time for all drive-through orders. (This was the information the operation's owners now want each day.)

Assume you were Loretta. Why do you think the owners are now requesting this new employee productivity report? How much do you believe the information contained in this report will help your labor cost control efforts, and your efforts to optimize guest service in the drive-through?

Key Terms

Salaried employee	Controllable	Productivity (worker)
Exempt employee	labor expense	Sales per labor hour
Job description	Non-controllable	Labor dollars per
Job specification	labor expense	guest served
Split-shift	Tip distribution program	Labor cost percentage
Fixed payroll	Tip sharing	Empowerment
Variable payroll	Tip pooling	(employee)

Operator's 10-Point Tactics for Success Checklist

Evaluate your need for, and the current status of, each of the following operational tactics. For those tactics you think are important, but not yet in place, develop an action plan for its implementation including who will be responsible for the tactic's completion and the target date by which it should be completed.

Tactic	Don't Agree (Not Done)	Agree (Done)	Agree (Not Done)	If Not Done Who Is Responsible?	Target Completion Date
1) Operator understands the relationship between profits and controlling labor costs.	____	____	____		
2) Operator understands the difference between fixed payroll and variable payroll.	____	____	____		
3) Operator understands the difference between controllable and non-controllable labor expenses.	____	____	____		
4) Operator can identify the requirements of the Fair Labor Standards Act (FLSA) that apply to their own business.	____	____	____		
5) Operator recognizes the importance of properly accounting for the wages of tipped employees.	____	____	____		
6) Operator knows how to calculate and assess the "Sales per Labor Dollar" measure of productivity.	____	____	____		
7) Operator knows how to calculate and assess the "Labor Dollars per Guest Served" measure of productivity.	____	____	____		
8) Operator knows how to calculate and assess the "Labor Cost Percentage" measure of productivity.	____	____	____		

(Continued)

Tactic	Don't Agree (Not Done)	Agree (Done)	Agree (Not Done)	If Not Done	
				Who Is Responsible?	Target Completion Date
9) Operator understands the various actions that can be taken to reduce fixed and variable labor-related expenses.	___	___	___		
10) Operator recognizes the impact employee empowerment can have on reducing labor costs and improving profits.	___	___	___		

10

Operating Budgets

What You Will Learn

1) The Importance of Operating Budgets
2) How to Create an Operating Budget
3) How to Compare Actual Operating Results to Planned Results

Operator's Brief

In this chapter, you will learn the importance of creating and properly monitoring an operating budget. The operating budget, or financial plan, is developed to help a business reach its future financial goals. The operating budget tells foodservice operators what must be achieved to meet predetermined cost and profit objectives.

To create operating budgets, operators must first consider:

1) Prior-period operating results (if an existing operation)
2) Assumptions made about the next period's operations
3) The operation's financial objectives

An effective operating budget estimates a business's future sales, expenses, and profits for a specific accounting period. To estimate future sales most accurately you must review historical records, and then consider any internal or external factors that may affect future revenue generation. The next step in the budgeting process is to estimate each fixed, variable, and mixed cost.

As you will learn, one important result of budget creation is the ability to make comparisons between budgeted and actual financial performance.

In this chapter, you will learn how to analyze your actual financial performance and compare it to the results you originally planned or budgeted. You will also learn how to use this information to take corrective action(s) or to modify the operating budget. Doing so will better ensure your operation can achieve all of its financial goals.

CHAPTER OUTLINE

The Importance of Operating Budgets

Just as the income statement (see Chapter 3) tells managerial accountants about past performance, the **operating budget,** or financial plan, is developed to help a business achieve its future goals. The operating budget tells foodservice operators what must be done to meet their predetermined cost and profit objectives.

An operating budget generally involves several activities:

Key Term

Operating budget: An estimate of the income, expenses, and profit for a business over a defined accounting period. Also referred to as a "financial plan."

1) Establishing realistic financial goals
2) Developing a budget (financial plan) to achieve those goals
3) Comparing actual operating results with budgeted results
4) Taking corrective action, if needed, to modify operational procedures and/or the financial plan

Preparing a budget and staying within it helps operators meet financial goals. Without this plan, operators must guess how much to spend and how much sales should be anticipated. Effective operators build their budgets, monitor them closely, and modify them when necessary to achieve desired results.

Types of Operating Budgets

One helpful way to consider the purpose of an operating budget is by its coverage (time frame). While an operating budget may be prepared for any time desired, operating budget lengths are typically considered to be one of the three types shown in Figure 10.1.

Regardless of the time frame, all operating budgets are developed by planning estimated revenues, expenses, and profits associated with operating a business.

Type of Operating Budget	Budget Characteristics
Long-range	Typically prepared for a period of up to five years. While not highly detailed, it provides financial views about long-term goals.
Annual	Typically prepared for one calendar or fiscal year or, in some cases, one season. The budget may consist of 12 months or 13 periods of 28 days each.
Achievement	Prepared for a limited time, often a month, a week, or even a day. It most often provides very current operating information and greatly assists in making current operating decisions.

Figure 10.1   Operating Budget Types

Advantages of Operating Budgets

The owners of a foodservice operation want to know what they should expect to earn from their investments, and a budget helps to project those earnings. Questions related to the amount of revenue likely generated, the amount of cash that should be available for bill payment or distribution as earnings, and the proper timing of major purchases can all be addressed in a properly developed budget. As a result, in organizations of all sizes, proper budgeting is an essential process to be carefully planned and implemented.

The advantages of preparing and using an operating budget are many and are summarized in Figure 10.2.

1) It is a way to analyze alternative courses of action and allows operators to examine alternatives before adopting a specific plan.
2) It requires operators to examine the facts about what should be achieved in desired profit levels.
3) It enables operators to define standards used to develop and enforce appropriate cost control systems.
4) It allows operators to anticipate and prepare for future business conditions.
5) It helps operators periodically carry out a self-evaluation of their business and its progress toward meeting financial objectives.
6) It is a communication channel that allows the operation's objectives to be passed along to stakeholders including owners, investors, managers, and staff.
7) It encourages those who participated in budget preparation to establish their own operating objectives, evaluation tactics, and tools.
8) It provides operators with reasonable estimates of future expense levels and serves as an important aid in determining appropriate selling prices.
9) It identifies time periods in which operational cash flows may need to be supplemented.
10) It communicates realistic financial performance expectations of managers to owners and investors.

Figure 10.2   Advantages of Preparing and Using an Operating Budget

One way to consider operational budgets is to compare them to an operation's income statement (see Chapter 3). Recall that the income statement details the actual revenue, expenses, and profits incurred in operating a business. The operating budget, then, is simply an operator's best estimate of all (or any portion of) a future income statement.

Find Out More

Some foodservice operators and managerial accountants responsible for budget preparation for one or more units find membership in the Hospitality Financial and Technology Professionals (HFTP®) to be helpful. Established in 1952, HFTP is an international, nonprofit association headquartered in Austin, Texas, with offices in the United Kingdom, Netherlands, and Dubai. HFTP is recognized as the professional group representing the finance and technology segments of the hospitality industry.

HFTP offers members continuing education courses including those addressing budgeting through its HFTP Academy. To find out more about the educational resources offered by this professional group and their "Certified Hospitality Accountant Professional (CHAE)" program, enter "HFTP Academy" in your favorite search engine and view the results.

Creating an Operating Budget

Some operators believe it is difficult to develop an operating budget, and they do not take the time to do so. Creating an operating budget, however, does not need to be a complex process.

Before operators can begin to develop a budget, they must understand the essentials required for its creation. If these are not addressed before the operating budget is developed, the budgeting process that follows is not likely to yield an accurate or helpful financial planning tool.

Before beginning creating an operating budget, developers need to have and understand the following:

1) Prior-period operating results (if an existing operation)
2) Assumptions made about the next period's operations
3) Knowledge of the organization's financial objectives

1) Prior-Period Operating Results
   The task of budgeting is easier when an operator knows the results from prior accounting periods. Experienced managerial accountants know that what

occurred in the past is often an indicator of what may occur in the future. The further back and the greater detail with which an operator can track historical revenues and expenses, the more accurate a future budget will be.

For example, if operators know the revenues and expenses for the past 50 Saturdays, they are better able to forecast this coming Saturday's revenue and expense budgets than if they have the data for only the last two Saturdays available.

When preparing a budget, **historical data** should always be considered along with the most recent data available.

For example, assume a foodservice operator knows that revenues have, on an annual average, increased 5% each month from the same period last year. However, in the last two months, the increase has been closer to zero. This may mean that the revenue increase

Key Term

Historical data: Information about an operation's past financial performance including revenue generated, expenses incurred, and profits (or losses) realized.

trend has slowed or stopped completely. Good operators modify historical trends by closely examining current conditions. In the scenario being analyzed, perhaps the operator should estimate next month's revenue increases to be closer to 0% than to 5%. For new operations that have not yet opened, however, historical data will not be available.

2) Assumptions about the Next Period's Operations

Evaluating future conditions and business activity is always a key part of developing an operating budget. Examples include the opening of new competitive restaurants in the immediate area, scheduled occurrences including local sporting events and concerts, and significant changes in operating hours. Local newspapers, trade or business associations, and Chambers of Commerce are possible sources of helpful information about changes in future demand.

If significant changes are planned for an operation such as new menu items, changes in operating hours, or the estimated impact of marketing efforts, assumptions about the impact of these actions become important. After these factors have been considered, assumptions regarding revenues and expenses may be made.

3) Knowledge of the Organization's Financial Objectives

An operation's financial objectives may consist of a specific profit target defined as a percent of revenue or a total dollar amount and specific financial and operational ratios (see Chapter 4) that should be achieved. Many financial objectives are determined by an operation's owner(s) based on desired return on investment (ROI), and the operating budget must address these goals.

The operating budget is a detailed plan that can be expressed by the basic formula:

Budgeted Revenue – Budgeted Expense = Budgeted Profit

The budgeted profit level an operation seeks can be achieved when the operation realizes its budgeted revenue levels and spends only what has been budgeted to generate the sales. If revenues fall short of forecast and/or if expenses are not reduced to match the shortfall, budgeted profit levels are not likely.

In a similar manner, if actual revenues exceed forecasted levels, expenses (variable and mixed) will also increase. If the increases are monitored carefully and are not excessive, increased profits should result. If, however, an operator allows actual expenses to exceed the levels required by the additional revenue, budgeted profits may not be achieved.

To illustrate the operating budget development process, consider Jake, the owner/operator of Jake's Restaurant. He is developing the operating budget for next year and has determined historical budget data essentials as shown below:

1) He has gathered his prior-period (prior year) operating results.
2) From information applicable to the area's economic conditions and his competition, he has made the following assumptions about next year's operations:
   - Total revenues received will increase by 4% primarily due to a 4% menu price increase.
   - Food and beverage costs will increase by 3% due to inflation affecting food and beverage product prices. As a result, his product expense (cost of sales) targets are a 35% food cost and a 16% beverage cost, for a combined food and beverage total cost of sales of 31.2%.
   - Management, staff wages, and benefits costs have a target of 35% of sales for total labor cost.
   - Prime costs will be 66.2% of sales.
   - All other controllable expenses will total no more than 12.4% of sales.
3) Jake's financial objectives for the restaurant are to earn profits (net income before income taxes) of at least 11.1% of sales and net income (after taxes) of 8.3% of sales.

Given these assumptions, Jake is now ready to create the 12 monthly revenue and expense forecasts needed to complete next year's operating budget.

Revenue Forecasts

Accurately forecasting an operation's revenues is critical because all forecasted expenses and profits will be based on revenue forecasts. Managerial accountants know that, in most cases, revenues should be estimated on a monthly (or weekly) basis. Then they can be combined to create the annual revenue budget. The reason: Many hospitality operations have seasonal revenue variations.

For example, a restaurant doing a significant amount of business in its outside dining patio area may generate reduced sales during times of the year when inclement weather makes outside dining less desirable for guests. Similarly, a restaurant operated in a ski resort town will generally be busier in the winter season rather than summer months.

Forecasting revenues is not an exact science. However, it can be made more accurate when operators:

✓ **Review historical records.** Operators begin the revenue forecasting process by reviewing their revenue records from previous years. When an operation has been open at least as long as the budget period being developed, its revenue history is extremely helpful in predicting future revenue levels.

✓ **Consider internal factors affecting revenues.** In this step, operators must consider any significant changes in the type, quantity, and direction of their marketing efforts. Other internal activities that can impact future revenues include those related to facility renovation that might affect dining capacity and when the operation may be disrupted because of renovations. In some operations, the number of hours to be opened or menu prices may change. Any internal initiation or change an operator believes will likely impact future revenues should be considered in this step.

✓ **Consider external factors affecting revenues.** There are numerous external issues that could affect an operation's revenue forecasts. These include planned competitors' openings (or closings), and other factors such road improvements or construction that disrupts normal traffic patterns. Other factors that may yield revenue forecast changes are forecasted economic upturns or downturns that affect potential guests to spend discretionary income for hospitality services.

Returning to the example of Jake's Restaurant and using September as the month he is now working on, Jake has reviewed last year's data for the month of September and found his sales were $192,308, with an average guest check of $12.02.

He considers his internal and external factors affecting revenues, and he has estimated a net 4% increase in revenues for the coming September. Jake then computes his revenue forecast for the upcoming September as:

$$\text{Sales Last Year} \times (1.00 + \% \text{ Increase Estimate}) = \text{Revenue Forecast}$$

Or

$$\$192,308 \times (1.00 + 0.04) = \$200,000$$

Using historical data, Jake knows that approximately 80% of his sales are from food, and 20% of his sales are from beverages. Thus, he estimates $160,000 ($200,000 × 0.80 = $160,000) for food sales and $40,000 ($200,000 × 0.20 = $40,000) for beverage sales.

Jake's guest check average (including food and beverages) for last September was $12.02. With a forecasted increase of 4% in selling prices, the forecast for this September's check average will be calculated as:

Guest Check Average Last Year $\times$ (1 + % Increase Estimate)
= Guest Check Average Forecast

Or

$12.02 \times (1.00 + 0.04) = $12.50

With forecasted sales of $200,000 and a forecasted guest check average of $12.50, Jake's forecasted number of guests would be 16,000 ($200,000 revenue ÷ $12.50 guest check average = 16,000 guests).

In most cases, it is not realistic to assume an operator can forecast their business's exact monthly revenue one year in advance. With practice, accurate historical sales data, and a realistic view of internal and external variables that will affect an operation's future revenue generation, however, many operators can attain operating budget forecasts that are easily and routinely within 5–10% (plus or minus) of actual results for a forecasted accounting period.

Technology at Work

Today's readily available restaurant sales forecasting tools help foodservice operation owners better estimate future volume and make more informed and accurate decisions about purchasing and staffing. These advanced technology tools, often interfaced with an operation's point-of-sale (POS) system, can use historical data to help foresee upcoming seasonality trends and predict how the trends will likely affect an operation's revenue generation. Since they assist in establishing realistic sales and profit goals, the result is better planning and preparation for the future.

To examine some offerings of companies that have developed software to assist in sales forecasting, enter "restaurant revenue forecasting software" in your favorite search engine and review the results.

Expense Forecasts

Operators must budget for each fixed, variable, and mixed cost (see Chapter 6) when they address individual expense categories on income statements. Fixed costs are simple to forecast because items such as rent, depreciation, and interest typically do not vary from month to month.

Variable costs, however, are directly related to the amount of revenue produced by a foodservice operation. For example, an operation that forecasts the sale of one hundred prime rib dinners on Friday night will likely have higher food and server (labor) costs than the operator in a similar facility that forecasts the sale of only 50 prime rib dinners. The reason: Variable expenses such as food, beverage, and labor costs are affected by sales levels.

Mixed costs, however, contain both fixed and variable cost components. For example, an operation will have a minimum number of employees (fixed costs) who must be scheduled to work even when business is slow. However, as the number of guests served increases, the operation will spend additional labor dollars (variable costs) to properly serve its guests.

Forecasting Fixed Costs

In most cases, budgeting fixed costs is easy. For example, Jake knows his monthly rent (occupancy) payments are $10,000. Creating the annual budget for "rent" expense is a simple matter of multiplication: $10,000 a month for 12 months yields a total annual occupancy cost of $120,000 ($10,000 × 12 months = $120,000). In this situation, since rent is a fixed cost, it remains unchanged regardless of the revenues generated by the restaurant. Note: Any anticipated increases in fixed costs for the coming year must be budgeted separately for each month of the year.

Forecasting Variable Costs

Variable costs increase or decrease as revenue volumes change. For example, consider the cost of linen napkins used at Jake's Restaurant. As more guests are served, more napkins will be used, and higher napkin expenses will be incurred because of laundry charges. The laundry charges for napkins in this situation are variable costs. The food Jake will purchase, the beverages he will serve, and a variety of other expenses are all variable costs that increase as the number of guests he serves increase.

Variable costs can be forecasted using targeted percentages or costs per unit (guests served). For example, food costs might be forecasted at a targeted food cost percentage. Similarly, an operator might forecast cleaning supplies as a percentage of total sales. When percentages are used, the sales forecast is multiplied by the target cost percentage to arrive at the forecasted cost.

In the case of Jake's Restaurant, a targeted food cost percentage of 35% and $160,000 in estimated food sales would yield the following forecasted cost of sales for food:

$$\text{Revenue Forecast} \times \text{Targeted Cost \%} = \text{Forecasted Cost}$$

Or

$$\$160,000 \times 0.35 = \$56,000$$

Jake will use this same approach and formula to estimate beverage costs and all other variable expenses.

Forecasting Mixed Costs

To create the expense portion of his operating budget, Jake must accurately forecast his mixed costs. One of the largest line-item costs in many foodservice operations is that of labor. It is a mixed cost because it includes hourly wages (variable costs), salaries (fixed costs), and employee benefits (mixed costs). Experienced operators know that, when labor costs are excessive, profits are reduced. As a result, this is an area of budgeting and cost control that is extremely important. Accurate budgets used to help control future labor costs can be precisely calculated using a three-step method.

Step 1: Determine Targeted Total Labor Dollars to Be Spent

In most cases, determining total labor costs are tied to the targeted or standard costs an operation desires. These standards or goals may be established by considering the historical performance of an operation, by referring to industry segment or company averages, or by considering the profit level targets of the business. In most cases, the standard will be developed by a consideration of each of these important factors.

When all relevant labor-related information has been considered, a labor standard (see Chapter 9) is used to determine the operating budget for total labor.

Jake has set his total labor cost standard to be 35% of total sales. Thus, with a $200,000 sales forecast for September, and a 35% labor cost percentage standard, the total amount to be budgeted for labor (salaries, wages, and employee benefits) would be calculated as:

$$\text{Revenue Forecast} \times \text{Labor Cost \% Standard} = \text{Forecasted Total Labor Cost}$$

Or

$$\$200,000 \times 0.35 = \$70,000$$

Step 2: Subtract the Cost of Employee Benefits

Employee benefits consist of costs associated with, or allocated to, payroll, and they must be paid by employers. Examples include items such as payroll taxes (e.g., mandated contributions to Social Security and worker's compensation plans) and voluntary benefit programs offered by the operation.

Key Term

Employee benefits: Any form of rewards or compensation provided to employees in addition to their base salaries and wages. Benefits offered to employees may be legally mandatory or voluntary.

The costs of these mandatory and voluntary benefit programs can be significant, and they include costs such as:

✓ Performance bonuses
✓ Health, dental, vision, and hearing insurance
✓ Life insurance
✓ Long-term disability insurance
✓ Employee meals
✓ Sick leave
✓ Paid holidays
✓ Paid vacation

Foodservice operators may offer all or some of these benefits; however, the applicable mandatory and voluntary payroll allocations must be subtracted from the labor dollars available to be spent when developing an accurate operating budget.

For Jake's Restaurant, assume that, historically, mandatory and voluntary benefits have accounted for 20% of the total labor costs incurred by the operation. The calculation required to determine the budgeted payroll allocation (employee benefits) amount would then be:

$$\text{Forecasted Labor Cost} \times \text{Employee Benefits \%} = \text{Budgeted Employee Benefits}$$

Or

$$\$70,000 \times 0.20 = \$14,000$$

In this example, the amount remaining for use in paying all operational salaries and hourly wages (budgeted payroll) would be computed as:

$$\text{Forecasted Labor Cost} - \text{Budgeted Employee Benefits} = \text{Budgeted Payroll}$$

Or

$$\$70,000 - \$14,000 = \$56,000$$

Step 3: Subtract Management (Fixed) Costs to Determine Staff (Variable) Costs

In most cases, fixed payroll remains unchanged from one pay period to the next unless an individual receiving fixed pay separates employment from the organization. Management labor costs are typically a fixed cost. Staff costs, alternatively, consist primarily of those dollars paid to hourly employees. Variable payroll is the amount that "varies" with changes in sales volume. The distinction between fixed and variable labor is an important one since managers may sometimes have little control over their fixed labor costs while exerting nearly 100% control over variable labor costs.

To determine the amount of money to be budgeted for (hourly) staff, Jake must first subtract the management (fixed) portion of his labor costs. This fixed labor component consists of all the operational salaries he will pay. These labor costs must be budgeted and subtracted from the total available for labor because the salary amounts will be paid regardless of sales volume.

To illustrate the budgeting process, assume Jake pays $18,000 in salaries each month. The amount to be budgeted for staff labor costs for September would be calculated as:

Budgeted Payroll – Management Costs = Budgeted Staff Costs

Or

$$\$56,000 - \$18,000 = \$38,000$$

Thus, $38,000 will be available to be paid to Jake's front-of-house and back-of-house hourly paid employees.

In most foodservice operations, managers who successfully create an operating budget for their food, beverages, and labor will have accounted for more than 50% of their total costs. All other expenses can be budgeted using the same methods for fixed, variable, and mixed costs. A successful operator will have separated all mixed costs into their fixed and variable components using the high/low method described in Chapter 6. This will make it easier to project changes in variable costs (and variable components of mixed costs) due to forecasted changes in sales.

By successfully forecasting his revenue and expenses, Jake can now develop the entire budget for his operations for September. Recall that Jake's assumptions and financial objectives were as follows:

✓ Total revenues received will increase by 4% primarily due to a 4% menu price increase.
✓ Food and beverage costs will increase by 3%. Targets are a 35% food cost and a 16% beverage cost, for a total combined food and beverage cost of sales of 31.2%.
✓ Management, staff wages, and benefits have a target of 35% of sales for total labor cost.
✓ All other controllable expenses will total no more than 12.4% of sales.
✓ Net income (after taxes) will be 8.3% of sales.

Based on this information, the resulting budget for Jake's Restaurant for September of the budget year is shown in Figure 10.3.

As can be seen in Figure 10.3, every fixed, variable, and mixed cost, considered individually, must be included in the operating budget. When that is completed, the result will be an operating budget that:

1) Is based upon a realistic revenue estimate.
2) Considers all known fixed, variable, and mixed costs.

Jake's Restaurant

Budget for September

Budgeted Number of Guests = 16,000

	Next Year	%
SALES		
Food	$160,000	80.0
Beverage	$ 40,000	20.0
Total Sales	$200,000	100.0
COST OF SALES		
Food	$ 56,000	35.0
Beverages	$ 6,400	16.0
Total Cost of Sales	$ 62,400	31.2
LABOR		
Management	$ 18,000	9.0
Staff	$ 38,000	19.0
Employee Benefits	$ 14,000	7.0
Total Labor	$ 70,000	35.0
PRIME COST	$132,400	66.2
OTHER CONTROLLABLE EXPENSES		
Direct Operating Expenses	$ 7,856	3.9
Music & Entertainment	$ 1,070	0.5
Marketing	$ 3,212	1.6
Utilities	$ 5,277	2.6
General & Administrative Expenses	$ 5,570	2.8
Repairs & Maintenance	$ 1,810	0.9
Total Other Controllable Expenses	$ 24,795	12.4
CONTROLLABLE INCOME	$ 42,805	21.4
NON-CONTROLLABLE EXPENSES		
Occupancy Costs	$ 10,000	5.0
Equipment Leases	$ —	0.0
Depreciation & Amortization	$ 3,400	1.7
Total Non-Controllable Expenses	$ 13,400	6.7
RESTAURANT OPERATING INCOME	$ 29,405	14.7
Interest Expense	$ 7,200	3.6
INCOME BEFORE INCOME TAXES	$ 22,205	11.1
Income Taxes	$ 5,551	2.8
NET INCOME	$ 16,654	8.3

Figure 10.3 Jake's Restaurant Operating Budget for September Next Year

3) Is intended to achieve the organization's financial goals.
4) Can be monitored to ensure adherence to the budget's guidelines.
5) May be modified when necessary.

Annual operating budgets are most often the compilation of monthly operating budgets. Therefore, operators may utilize their monthly operating budgets to create annual budgets and to monitor weekly (or even daily) versions of their overall operating budgets.

What Would You Do? 10.1

"I just don't' know," said Trishauna, "it might make an impact, but it might not make a very big one. I think we'll just have to wait and see."

Trishauna was talking to Joshua, her partner in the Wide Awake Café, a coffee and pastry shop located adjacent to the State University campus.

Trishauna and Josh were preparing next year's operating budget. They had just begun the process and were forecasting their next year's revenues. Both Trishauna and Joshua were aware that a major coffee shop chain just announced plans to open a new shop within one block of the Wide Awake Café.

"Well," said Joshua, "I think our clientele is loyal. I know our new competitor will generate some business, but I'm not sure how much of that business will come from us versus other coffee shops in the area."

"Exactly," replied Trishauna. "That's why I think it'll be a real challenge for us to forecast our revenue for next year."

Assume you were the owners of the Wide Awake Café. How important will it be for you to consider external influences such as the opening of a new competitor as you create your next year's revenue budget? What could a likely result be if you did not make such considerations?

Monitoring the Operating Budget

An operating budget details a plan for future financial activities. In many cases, an operator's forecast of future results will be reasonably accurate, and in other cases, it will not be as accurate. For example, revenue may not reach forecasted levels, expenses may exceed estimates, and internal or external factors not considered when the operating budget was prepared may negatively or positively impact financial performance.

An operation's budget will have little value if management does not utilize it. In general, the operating budget should be regularly monitored in each key area:

✓ Revenue
✓ Expenses
✓ Profit

Revenue

If sales should fall below projected levels, the impact on profit can be substantial, and it may be impossible to meet profit goals. If revenues consistently exceed projections, the overall budget must be modified or, ultimately, the expenses associated with these increased sales will soon exceed their budgeted amounts. Effective operators compare their actual sales to that which they have projected on a regular basis.

Expenses

Foodservice operators must be careful to monitor their operating expenses because costs that are too high or too low may be cause for concern. Just as it is not possible to estimate future sales volumes perfectly, it typically is also not possible to estimate future expenses perfectly, and some expenses will vary as sales volumes increase or decrease.

As business conditions change, revisions in the operating budget are to be expected. This is true because operating budgets are based on a specific set of assumptions and, as the assumptions change, so too will the accuracy of the operating budget produced from those assumptions.

To illustrate, assume an operator budgeted $1,000 in January for snow removal from the parking lot attached to a restaurant operating in upper New York State. If unusually severe weather causes the operator to spend $2,000 for snow removal in January, the assumption (normal levels of snowfall) was incorrect, and the original budget will be incorrect as well.

Profit

An operation's budgeted profit must be realized if the operation is to provide adequate returns for its owner. To illustrate, consider the case of Jessica, the manager of a foodservice establishment with excellent sales but below-budgeted profits. For this year, Jessica budgeted a 5% profit on $2,000,000 of sales; therefore, $100,000 profit ($2,000,000 × 0.05 = $100,000) was anticipated.

At year's end, Jessica achieved her sales goal, but doing so generated only $50,000 profit, or 2.5% of sales ($50,000 ÷ $2,000,000 = 2.5%). If this operation's owners feel that $50,000 is an adequate return for their investment and risk, Jessica's services may be retained. If they do not, she may lose her position, even though she operates a "profitable" restaurant. The reason: Management's task is not merely to generate a profit, but rather to generate the profit that was planned.

Comparing Planned Results to Actual Results

One important task of a foodservice operator is to optimize profitability by analyzing the differences between the planned for (budgeted) results and the actual operating results. To do this effectively, operators must receive timely income statements that accurately detail the actual operating results. When they do, these actual results can then be compared to the operating budget for the same accounting period. The difference between planned results and actual results is called budget **variance.**

The basic formula used to calculate a budget variance is:

$$\text{Actual Results} - \text{Budgeted Results} = \text{Variance}$$

A variance may be expressed in either dollar or percentage terms, and it can be either positive (favorable) or negative (unfavorable). A **favorable variance** occurs when the variance is an improvement on the budget (revenues are higher or expenses are lower). An **unfavorable variance** occurs when actual results do not meet budget expectations (revenues are lower or expenses are higher).

For example, if the budget for snow removal services is $1,000 for a given month, but the actual expenditure for those services is $1,250, the variance is calculated as:

$$\text{Actual Results} - \text{Budgeted Results} = \text{Variance}$$

Or

$$\$1,250 - \$1,000 = \$250$$

In this example, the variance may be expressed as a dollar amount ($250) or as a percentage of the original budget. The computation for a percentage variance is:

$$\frac{\text{Variance}}{\text{Budgeted Results}} = \text{Percentage Variance}$$

Or

$$\frac{\$250}{\$1,000} = 0.25\,(25\%)$$

In this example, the variance is unfavorable to the operation because the actual expense is higher than the budgeted expense. In business, a variance can also be considered as either significant or insignificant. It is the operator's task to identify significant variances between budgeted and actual operating results.

A **significant variance** may be defined several ways; however, a common definition is that a significant variance is any difference in dollars or percentage between budgeted and actual operating results that warrants further investigation. Significant variance is an important concept because not all variances need to be investigated.

Key Term

Significant variance: A variance that requires immediate management attention.

For example, assume that, at the beginning of a year, a foodservice operator prepares an annual (12 months) operating budget that forecasts the operation's December utility bill will be $6,000. When, 12 months later, the December bill arrives, it totals $6,420, and therefore it is $420 over budget ($6,420 actual − $6,000 budgeted = $420 variance).

Given the amount of the bill ($6,420) and the difficulty of accurately estimating utility expenses one year in advance, a difference of only $420 or 7% ($420 ÷ $6,000 = 0.07) probably does *not* represent a significant variance from the operation's budget.

Alternatively, assume that the same operation had estimated office supplies usage at $100 for that same month, but the actual cost of supplies was $520. Again, the difference between the budgeted expense and actual expense is $420. The office supplies variance, however, represents a very significant difference of 420% ($420 ÷ $100 = 4.20 or 420%) between planned and actual results, and it should probably be thoroughly investigated.

Managerial accountants decide what represents a significant variance based on their knowledge of the specific operation, and their own company policies and procedures. Small percentage differences can be important if they represent large dollar amounts. Similarly, small dollar amounts can be significant if they represent large percentage differences from planned results. Variations from budgeted results can occur in revenues, expenses, and profits. Managers can monitor all these areas using a four-step operating budget monitoring process.

Step 1 Compare actual results to the operating budget.
Step 2 Identify significant variances.
Step 3 Determine causes of variances.
Step 4 Take corrective action or modify the operating budget.

In Step 1, an operator reviews the income statement and operating budget data for a specified accounting period. In Steps 2 and 3, actual operating results are compared to the budget and significant variances, if any, are identified and analyzed. Finally, in Step 4, corrective action is taken to reduce or eliminate unfavorable variances, or if it is appropriate to do so, the budget is modified to reflect new realities confronting the business.

In most cases, operators should compare their actual results to their operating budget results in each of the income statement's three major sections of sales, expense, and profits. To illustrate the process, consider again the operating budget for Jake's Restaurant. Figure 10.4 shows Jake's original operating budget

Jake's Restaurant

Budget versus Actual Comparison for September

	Budgeted Number of Guests = 16,000		Actual Number of Guests = 15,500	
	Budget	**%**	**Actual**	**%**
SALES				
Food	$160,000	80.0%	$150,750	79.3%
Beverage	$ 40,000	20.0%	$ 39,250	20.7%
Total Sales	$200,000	100.0%	$190,000	100%
COST OF SALES				
Food	$ 56,000	35.0%	$ 53,800	35.7%
Beverages	$ 6,400	16.0%	$ 6,300	16.1%
Total Cost of Sales	$ 62,400	31.2%	$ 60,100	31.6%
LABOR				
Management	$ 18,000	9.0%	$ 18,000	9.5%
Staff	$ 38,000	19.0%	$ 37,800	19.9%
Employee Benefits	$ 14,000	7.0%	$ 11,160	5.9%
Total Labor	$ 70,000	35.0%	$ 66,960	35.2%
PRIME COST	$132,400	66.2%	$127,060	66.9%
OTHER CONTROLLABLE EXPENSES				
Direct Operating Expenses	$ 7,856	3.9%	$ 7,750	4.1%
Music & Entertainment	$ 1,070	0.5%	$ 1,070	0.6%
Marketing	$ 3,212	1.6%	$ 1,350	0.7%
Utilities	$ 5,277	2.6%	$ 5,195	2.7%
General & Administrative Expenses	$ 5,570	2.8%	$ 5,455	2.9%
Repairs & Maintenance	$ 1,810	0.9%	$ 1,925	1.0%
Total Other Controllable Expenses	$ 24,795	12.4%	$ 22,745	12.0%
CONTROLLABLE INCOME	$ 42,805	21.4%	$ 40,195	21.2%
NON-CONTROLLABLE EXPENSES				
Occupancy Costs	$ 10,000	5.0%	$ 10,000	5.3%
Equipment Leases	$ –	0.0%	$ –	0.0%
Depreciation & Amortization	$ 3,400	1.7%	$ 3,400	1.8%
Total Non-Controllable Expenses	$ 13,400	6.7%	$ 13,400	7.1%
RESTAURANT OPERATING INCOME	$ 29,405	14.7%	$ 26,795	14.1%
Interest Expense	$ 7,200	3.6%	$ 7,200	3.8%
INCOME BEFORE INCOME TAXES	$ 22,205	11.1%	$ 19,595	10.3%
Income Taxes	$ 5,551	2.8%	$ 4,899	2.6%
NET INCOME	$ 16,654	8.3%	$ 14,696	7.7%

Figure 10.4 Jake's Restaurant Budget versus Actual Comparison for September

(see Figure 10.3) and his actual operating results in dollars and percentages of sales for the month of September.

If Jake properly monitors his budget, sales is the first area he will examine when comparing his actual results to his budgeted results. This is true because, if sales fall significantly below projected levels, there will likely be a significant negative impact on profit goals.

Secondly, when sales vary from projections, variable costs also fluctuate. In cases where sales are lower than budget projections, variable costs should be less than budgeted. In addition, when actual sales fall short of budgeted levels, fixed and mixed expenses, such as rent and labor, incurred by the operation will represent a larger-than-originally-budgeted *percentage* of total sales. Alternatively, when actual sales exceed the budget, total variable expenses will increase, and fixed and mixed expenses incurred should, if properly managed, represent a smaller-than-budgeted percentage of total revenue.

A close examination of Figure 10.4 shows that Jake has experienced a shortfall in both food and beverage revenue when compared to his operating budget. One revenue problem Jake faced in September is that he budgeted for 16,000 guests and only served 15,500 guests. Also, Jake budgeted for a $12.50 average guest check ($200,000 budgeted sales/16,000 budgeted guests = $12.50 budgeted guest check average) but achieved a guest check average of only $12.26 ($190,000 actual sales/15,500 actual guests = $12.26 actual guest check average).

If sales consistently fall short of forecasts, Jake must evaluate all aspects of his entire operation to identify and correct the revenue shortfalls. Foodservice operations that consistently fall short of their sales projections must also evaluate the validity of the primary assumptions used to produce the sales portion of their operating budgets.

Find Out More

To achieve desired sales levels, foodservice operators must have an effective marketing plan in place. Increasingly, an operation's marketing efforts must include both traditional approaches and newer Internet-based approaches.

While there are many publications addressing general marketing strategies for businesses, few publications exclusively address the marketing needs of foodservice operations.

One of the best and most up-to-date marketing resources available to foodservice operators, and one that exclusively addresses the on-site and on-line marketing of foodservice operations is "*Marketing in Foodservice Operations*" by David K Hayes and Jack D Ninemeier.

To learn more about the content and availability of this extremely valuable new publication, enter "*Marketing in Foodservice Operations*" in your favorite search engine and review the results.

Expense Analysis

Identifying significant variances in expenses is a critical part of the budget monitoring process because many types of operating expenses are controllable. Some variation between budgeted and actual costs can be expected because most variable operating expenses vary with sales levels that cannot be predicted perfectly. The variances that occur can, however, tell operators a great deal about operational efficiencies, and experienced operators know that a key to ensuring profitability is to properly examine and manage controllable costs.

As shown in Figure 10.4, Jake's cost of food in dollars were lower than budgeted ($56,000 budgeted vs. $53,800 actual) as he would expect given his food sales shortfall, but his cost of sales percentage was over-budget (35.0% budgeted vs. 35.7% actual), an indication that his kitchen was not as cost-efficient as he would have liked. The same was true in his beverage sales category.

Similarly, Jake's total dollar amount spent for labor (management, staff, and employee benefits) was lower than budgeted ($70,000 budgeted vs. $66,960 actual), but his reduced sales level resulted in a slight increase in total labor percentage (35.0% budgeted vs. 35.2% actual). Note also in Figure 10.4 that Jake's fixed costs (e.g., occupancy, depreciation, and interest) did not vary in dollar amount. However, because revenues did not reach their budgeted levels, these too represent a higher actual cost percentage of sales because the fixed dollars were spread over a smaller revenue base.

The marketing expense category in Figure 10.4 is a line item that illustrates well the need to compare budgeted expense to actual expense. Jake's September marketing budget was $3,212. His actual expense in that category was $1,350. While this variance might initially seem to be positive (because less was spent on this expense than was previously budgeted), it is likely that some of Jake's shortfall in revenue came from the fact that he reduced his marketing expense. Experienced foodservice operators know that savings achieved in marketing costs often result in no savings at all!

Note also that Controllable Income in Figure 10.4 was budgeted at 21.4% of total revenue, yet the actual results were lower, at 21.2%. The difference in the dollar amount of Controllable Income achieved was $2,610 ($42,805 budgeted vs. $40,195 actual = $2,610 variance). Jake must decide if this constitutes a "significant" variance, and if so, it must be analyzed using the four-step "Operating Budget Monitoring Process" presented earlier in this chapter.

Profit (Net Income) Analysis

A foodservice operation's actual level of profit is measured either in dollars, percentages, or both, and is the most critical number that most operators evaluate. Returning to Figure 10.4, it is easy to see that Jake's actual net income for the month was $14,696 or 7.7% of total sales. This is less than the $16,654 (8.3% of total sales) that was forecast in the operating budget. This reduced profit level can be tied directly to Jake's lower-than-expected sales.

When a business is unable to meet its top-line sales forecasts, it typically means the budget was ineffectively developed, internal/external conditions have changed, and/or that the operation's marketing efforts were not effective. Regardless of the cause, when sales do not reach forecasted levels, corrective action may be needed to prevent even more serious problems including significant future profit erosion.

Foodservice profits (net income) are routinely reported on the income statement in both dollars and percentages, and foodservice operators may disagree about the best way to evaluate their profitability. One frequently heard comment in the hospitality industry is, "You bank dollars, not percentages!" To better understand this statement, note in Figure 10.4 that Jake achieved net income that was a lower percentage than the operating budget forecast (7.7% actual net income vs. 8.3% budgeted net income).

However, consider the hypothetical results that would be obtained if Jake's actual net income percentage in September had been 8.5%. This would have been a *higher* percentage than his operating budget predicted (8.5% actual vs. 8.3% budgeted).

In this hypothetical scenario, the $16,150 total dollars of profit that would be generated ($190,000 actual sales × 8.5% net income = $16,150) would still be *less* than the initially budgeted 8.3% profit of $16,654. This, then, is the reasoning behind the statement, "You bank dollars, not percentages!"

Modifying the Operating Budget

The best managerial accountants know an operating budget should be an active and potentially evolving document. The budget should be regularly reviewed and, when necessary, modified as new and better information replaces that available when the original operating budget was developed. This is especially true when new information significantly (and perhaps permanently) affects the sales and expense assumptions used to create the budget. The operating budget should be reviewed anytime it is believed the assumptions upon which it is based are no longer valid.

To illustrate, assume a foodservice operator employs 25 full-time employees. Each employee is covered under the operation's group health insurance policy. Last year, the operator agreed to pay 50% of each employee's insurance cost and, as a result, paid $300 per month for every full-time employee. The total cost of the insurance contribution each month was $7,500 (25 employees × $300 per employee = $7,500).

When this year's budget was developed, the operator assumed a 10% increase in health insurance premiums. If, later in the year, it is determined that premiums will actually be increased by 20%, employee benefit costs will be much greater than projected in the original operating budget. This operator now faces several choices:

✓ Modify the budget.
✓ Reduce the amount contributed per employee to stay within the budget.
✓ Change (reduce) health insurance benefits/coverage to lower the premiums that will be charged to stay within the original costs allocated in the operating budget.

Regardless of the operator's decision, the operating budget, if affected, must be modified. There are situations in which an operating budget should be legitimately modified; however, an operating budget should never be modified simply to compensate for management inefficiencies.

To illustrate, assume, for example, that a total labor percentage of 25% is realistic and achievable for a specific operation. The operation's managers, however, consistently achieve budgeted sales levels but just as consistently greatly exceed the labor cost percentage targets established by the operating budget. As this occurs, resulting profits are less than projected. In this case, the labor cost portion of the operating budget should not be increased (nor should menu prices be increased!) simply to mask management's inefficiencies in controlling costs in this area. Instead, if the goal of a 25% labor cost is indeed reasonable and achievable, then that operation's managers must correct the problem.

Properly prepared operating budgets are designed to be achieved, and foodservice operators must do their best to ensure this occurs. There are cases, however, when operating budgets must be modified or lose their ability to assist managers with decision-making responsibilities. The following situations are examples of those that, if unknown and not considered when the original budget was developed, may require operators to consider modifying the existing operating budget:

✓ Additions or subtractions from offerings that materially affect revenue generation (for example, reduced or increased operating hours)
✓ The opening of a direct competitor
✓ The closing of a direct competitor
✓ A significant and long-term or permanent increase or decrease in the price of major cost items
✓ Franchisor-mandated operating standard changes that directly affect (increase) costs
✓ Significant and unanticipated increases in fixed expenses such as mortgage payments (for example, a loan repayment plan tied to a variable interest rate), insurance, or property taxes
✓ A manager or key employee change that significantly alters the skill level of the facility's operating team

✓ Natural disasters such as floods, hurricanes, or severe weather that significantly affects forecasted revenues

✓ Changes in financial statement formats or expense assignment policies

✓ Changes in the investment return expectations of the operation's owners

This chapter addressed the importance of operating budgets in the management of a foodservice operation. Budgeting is an effective way to control expenses, but if funds are to be available to pay legitimate expenses it is also important to safeguard the revenue a foodservice operation generates. Effective revenue control procedures begin at the time of guest payment and continue when a deposit is made in the operation's bank account. The proper control of revenue is so important it is the sole topic of the next chapter.

Technology at Work

The preparation and monitoring of operating budgets in the foodservice industry can be a time-consuming process.

Today, however, there are a variety of software programs available, in both PC and Mac formats, to help the owners of foodservice businesses develop and monitor their operating budgets.

To review the features of some of these helpful tools, enter "budgeting software for restaurants" in your favorite search engine and view the results.

What Would You Do? 10.2

"Well, it just makes sense that we start with how much profit we need to make and then estimate the sales and expenses required to make that profit level. Otherwise, why are we buying the food truck?" said Ralphie.

Ralphie was talking to Catriona, his partner in a new joint venture in which Ralphie and Catriona were sharing the cost of buying a food truck that would feature English style Fish and Chips. They had agreed to share the truck's cost and operating profits on a 50–50 basis, and they were now preparing their first year's operating budget.

"I don't know if I agree with that," said Catriona. "It seems to me we need to estimate our realistic sales levels first and then our operating costs to see what profit we're likely to generate in the first year. Just because we want to make a certain amount of profit doesn't mean that we will!"

Assume you were Ralphie. Why do you think it makes sense to focus on profits first when preparing an operating budget? Assume you were Catriona. Why do you think it is important to focus on estimated sales levels when preparing a first year's operating budget?

Key Terms

Operating budget	Variance (budget)	Unfavorable variance
Historical data	Favorable variance	Significant variance
Employee benefits		

Operator's 10-Point Tactics for Success Checklist

Evaluate your need for, and the current status of, each of the following operational tactics. For those tactics you think are important, but not yet in place, develop an action plan for its implementation including who will be responsible for the tactic's completion and the target date by which it should be completed.

				If Not Done	
Tactic	Don't Agree (Not Done)	Agree (Done)	Agree (Not Done)	Who Is Responsible?	Target Completion Date
1) Operator understands the importance to profits of developing an accurate operating budget.	——	——	——		
2) Operator can identify the unique purposes of each of the three different types of operating budgets.	——	——	——		
3) Operator can state the advantages that will accrue to them when they utilize well-prepared and detailed operating budgets.	——	——	——		
4) Operator recognizes the importance of assessing prior-period operating results when preparing an operating budget.	——	——	——		
5) Operator recognizes the importance of utilizing realistic assumptions about the next period's operations when preparing an operating budget.	——	——	——		

Tactic	Don't Agree (Not Done)	Agree (Done)	Agree (Not Done)	If Not Done	
				Who Is Responsible?	Target Completion Date
6) Operator recognizes the importance of understanding their organization's financial objectives when preparing an operating budget.	——	——	——		
7) Operator knows how to create realistic sales forecasts when preparing an operating budget.	——	——	——		
8) Operator knows how to create realistic fixed, variable, and mixed cost expense forecasts when preparing an operating budget.	——	——	——		
9) Operator has a system in place for identifying significant variances between actual operating results and planned results.	——	——	——		
10) Operator recognizes the importance of corrective action or budget modification when actual operating results vary significantly from planned results.	——	——	——		

11

Cash and Revenue Control

What You Will Learn

1) How to Identify External and Internal Threats to Revenue
2) How to Create Countermeasures to Combat Revenue Theft
3) How to Establish and Monitor an Effective Revenue Security System

Operator's Brief

One of a foodservice operator's most important responsibilities is to protect business assets from theft and fraud. Foodservices can be an easy victim of these crimes, so control systems maintaining asset security are essential.

In this chapter, you will learn about effective revenue control systems and how their components are developed, implemented, and monitored. A foodservice operation's cash assets are generated by guests' sales. Unfortunately, cash can be tempting to dishonest individuals, and its security is subject to external and internal threats that must be controlled.

In this chapter, you will learn important procedures for safeguarding cash from when it is received from guests until it is deposited in an operation's bank account. There are several objectives and useful strategies important as revenue security programs are developed, and this chapter explains them.

As revenue security is implemented and monitored, operators must pay special attention to five key activities. Each is detailed in this chapter, and you will learn why it is necessary to properly verify your own operation's:

Product Issues
Guest Charges
Sales Receipts
Sales Deposits
Accounts Payable (AP)

CHAPTER OUTLINE

The Importance of Revenue Control
    External Threats to Revenue
    Internal Threats to Revenue
Developing a Revenue Security Program
    Objectives of Internal Revenue Control
    Elements of Internal Revenue Control Systems
Implementing and Monitoring a Revenue Security Program
    Verification of Product Issues
    Verification of Guest Charges
    Verification of Sales Receipts
    Verification of Sales Deposits
    Verification of Accounts Payable (AP)

The Importance of Revenue Control

Foodservice operators need an effective system to control cash revenues. These controls are typically important because many foodservice operations have hundreds (or more!) daily cash sales transactions, and more than one employee typically handles the cash.

Fraud and **embezzlement** can occur all too frequently in some foodservice operations. Fortunately, there are several ways operators can reduce opportunities for these thefts to occur.

In most cases, there are three primary factors necessary for fraud and employee embezzlement to occur:

1) Need: Economic or psychological motives can encourage employees to steal. Some operators attempt to study their employees' lifestyles to determine whether selected staff members may steal. However, a better approach is to develop systems that reduce theft opportunities and that make it difficult for employees to try and "beat the system."

2) Failure of conscience: Sometimes thieves rationalize stealing to justify taking someone else's property. Some foodservice staff might believe that they deserve more money because they are paid too little, and "stealing a little is okay because the business can afford it."

Key Term

Fraud: Deceitful conduct to manipulate someone into giving up something of value by (a) presenting as true something known by the fraudulent party to be false or (b) by concealing a fact from someone that may have saved him or her from being cheated.

Key Term

Embezzlement: A crime in which a person or entity intentionally misappropriates assets that have been entrusted to them.

3) Opportunity: Some otherwise honest employees may be tempted to commit theft when given the opportunity to do so. Foodservice operators can discourage theft by implementing systems and procedures that minimize opportunities for employees to steal.

Of the three factors noted above, foodservice operators have the most influence over theft opportunities. Therefore, cash and revenue control procedures should be designed to minimize these opportunities.

Most foodservice operations have some general operating characteristics that make them more vulnerable to theft than many other businesses:

✓ Large numbers of individual cash transactions
✓ Items of relatively high value that are commonly used or available
✓ Availability of products employees must otherwise buy

Increasingly, foodservice guests use electronic payment cards rather than cash when paying their bills. However, cash is often exchanged between guests and employees. Cash banks made available to cashiers can create potential problems as well.

The use of relatively unskilled employees in low-paying positions with little social status may also contribute to high employee turnover rates. These, in turn, can influence the internal control environment. Many foodservice operations are small, and even large operations may be organized into several small revenue outlets such as individual dining rooms, multiple bars, and perhaps take-out stations. These can create strong needs for effective internal control systems that address both external and internal threats to revenue security.

External Threats to Revenue Security

Foodservice guests can be threats to an operation's revenue, and an operation can lose sales revenue because some guests want to defraud it. This activity can take a variety of forms including when guests **walk** (skip) without paying their bill. This type of theft is less likely in a quick-service restaurant (QSR) or fast casual operation because payment is typically collected before, or at the same time, guests receive their menu items.

Key Term

Walk (bill): Guest theft that occurs when a guest consumes products that were ordered and leaves without paying the bill. Also referred to as a "skip."

In cases where a guest is in a busy table-service restaurant's dining room, it is easier for one or more persons in a dining party to leave while the server is busy with other guests. In fact, it is sometimes easy for a guest or entire party to leave without settling their bill unless all staff members are vigilant. To help reduce this type of guest theft, implementation of the steps in Figure 11.1 is helpful.

Another form of external theft can be used by the quick-change artist: A guest who attempts to confuse a cashier to give the guest excessive change. For example, a guest who should have received $5 in change may use a confusing routine to secure $15. To prevent this from happening, operators must train cashiers and instruct them to notify management immediately if there is suspicion of attempted fraud from quick-change routines.

Internal Threats to Revenue Security

Most foodservice employees are honest, but some are not. In addition to protecting revenue from unscrupulous guests, operators must also be aware of employees who attempt to steal revenue from their properties.

Cash is the most readily usable asset in a foodservice operation, and it is a major target for dishonest employees. In general, theft from service personnel does not occur by removing large sums of cash at one time, because this would be too easy

---

1) If guests order and consume their food before payment, servers can present the bill promptly when guests finish eating.
2) If the operation has a cashier in a central location in the dining area, he or she should be available and visible at all times.
3) If each server collects his or her own guest charges, he or she should return to the table promptly after presenting the guest's bill for payment.
4) Train employees to observe all exit doors near restrooms or other areas of the facility that may provide opportunities to exit dining areas without being easily seen.
5) If an employee sees a guest attempting to leave without paying the bill, he or she should notify the manager immediately.
6) When approaching someone who has left without paying the bill, the manager should ask if the guest has inadvertently "forgotten" to pay. (In most cases, the guest will then pay the bill.)
7) If a guest still refuses to pay, or flees the scene, the manager should make a written note of the following:
   a) Number of guests involved
   b) Amount of bill
   c) Physical description of guest(s)
   d) Vehicle description, if applicable, with license plate number, if possible
   e) Time and date of the incident
   f) Name of the server(s) who served the guest
8) If the guest is successful in fleeing the scene, police should be notified. Note: In no case should staff members, supervisors, or managers attempt to physically detain the guest. The reason: Liability is involved if an employee or guest is hurt, and this may be far greater than the value of a skipped food and beverage bill.

Figure 11.1   Steps to Reduce Guest Walks, or Skips

for managers to detect. Rather, service personnel may use numerous methods of removing a small amount of money at different times.

One common theft technique used by servers involves failure to record a guest's order in the operation's point-of-sale (POS) system. Instead, the server charges the guest and keeps the revenue from the sale. To reduce this possibility, managers must insist that all sales be recorded in the POS system as a tactic that matches products sold to revenue received.

The POS system assigns a unique transaction number to the sale just as numbered guest checks did before introduction of POS systems. A sale's transaction number (or a numbered guest check) is an electronic or handwritten record of what the guest purchased and how much the guest was charged for the item(s).

The use of electronic guest checks produced by the POS system is standard in the foodservice industry. A rule for all food and beverage production personnel is that no food or beverage item should be issued to a server unless the server first records the sale in the POS system. The items ordered are then displayed in the operation's production area. In some systems, the order may even be printed in the production area.

Increasingly, this method of entering guest orders can be accomplished in one step instead of two. Handheld and wireless at-the-table order-entry devices now allow servers (and guests in some cases!) to enter orders directly into an operation's POS system. This direct data entry system is fast, and it eliminates mistakes made when transferring handwritten guest orders to the POS system.

Regardless of how the order is created, kitchen and bar personnel should not issue any products to the server without this uniquely numbered transaction. When the guest wants to leave, the cashier retrieves the transaction and prepares a bill for payment by the guest. The bill includes the charges for all items ordered by the guest plus any service charges and taxes due, and the guest then pays the bill.

Another method of employee theft involves entering sales but failing to collect payment. To prevent this theft, management must have systems in place to identify **open checks** during and after each server's work shift.

Unless all open checks are ultimately presented to guests for payment and closed out, the value of menu items issued will not equal the money collected for those items.

The totals of all transactions entered into the POS system during a predetermined time are electronically tallied, and managers can compare sales recorded by the POS system with the money in the cash drawer.

Key Term

Open check: A guest check that initially authorizes product issues from the kitchen or bar, but has not been collected for (closed) and, therefore, has not been added to the operation's sales total.

For example, a cashier working a shift from 7:00 a.m. to 3:00 p.m. might have recorded $1,000 in sales (including taxes) during that time. If that were the case and, if no errors in handling change occurred, the cash drawer should contain the $1,000 in sales revenue (in addition to the amount in the drawer at the beginning of the shift). If the drawer contains less than $1,000, it is said to be **short**; if it contains more than $1,000, it is said to be **over**.

Cashiers rarely steal large sums of cash from the cash drawer because this type of theft is easily detected. Managers should implement a policy to help ensure any cash shortages or overages will be investigated. Some managers believe that only cash shortages, not overages, need to be monitored, and this is not true.

Key Term

Short (cash bank): A cashier bank that, based on actual sales, contains less than the amount it should contain.

Key Term

Over (cash bank): A cashier bank that, based on actual sales, contains more than the amount it should contain.

Consistent cash shortages may be an indication of employee theft or carelessness and must be investigated. Cash overages, too, may result from sophisticated theft by the cashier. For example, assume a cashier defrauds an operation by removing and keeping $18.00 from a cash drawer but falsely reduces actual sales records by $20.00. The result is a $2.00 cash "overage!"

Even if operators implement controls to make internal theft difficult, the possibility of significant fraud still exists. Consequently, some companies protect themselves from employee dishonesty by **bonding** their employees. Bonding is a matter of management purchasing an insurance policy against the possibility that an employee(s) will steal.

Key Term

Bonding (employee): Protection from loss resulting from a crime event involving employee dishonesty and theft such as loss of money, securities, and other property.

When bonded, an employer can be covered for the loss of money or other property sustained through dishonest acts. Bonding can cover many acts including larceny, theft, embezzlement, forgery, misappropriation, or other fraudulent or dishonest acts committed by an employee alone or in collusion with others. Essentially, a business can select from several bonding options:

Individual—covers one employee (e.g., a foodservice operation's bookkeeper)
Position—covers all employees in a specific position (e.g., all bartenders or all cashiers)
Blanket—covers all employees

If an employee has been bonded, and an operation can determine that he or she was involved in the theft of a specific amount of money, the operation will be

1) Omits recording the guest's order and keeps the money the guest pays
2) Voids a sale in the POS system but keeps the money the guest paid
3) Enters another server's password in the POS system and keeps the money
4) Fails to finalize a sale (keeps a check open) and keeps the money
5) Charges guests for items not purchased and then keeps the overcharge
6) Changes the totals on payment card charges after the guest has left
7) Enters additional payment card charges and pockets the cash difference
8) Incorrectly adds legitimate charges to create a higher-than-appropriate total with the intent of keeping the overcharge
9) Purposely shortchanges guests when giving back change with the intent of keeping the extra change
10) Charges higher-than-authorized prices for products or services, records the proper price, and keeps the overcharge
11) Adds a coupon to the cash drawer and simultaneously removes sales revenue equal to the coupon's value
12) Declares a transaction to be complimentary (comped) after the guest has paid the bill
13) Engages in collusion between two or more employees to defraud the operation

Figure 11.2 Common Methods of Theft by Service Employees

reimbursed for all or part of the loss by the bonding company. Although bonding will not eliminate theft, it is a relatively inexpensive way to help ensure an operation is protected from theft by employees who handle cash or other forms of operating revenue. Note: The bonding company will likely require detailed background information on employees before bonding them, and this is also an excellent preemployment check to verify an employee's track record in prior jobs.

Operators must recognize that even good revenue control systems present the opportunity for theft if management is not vigilant, and this is especially so if two or more service employees work together to defraud the operation (collusion). Figure 11.2 identifies some common methods of theft involving foodservice employees.

The scenarios addressed in Figure 11.2 do not include all possible methods of revenue loss. However, operators must use an effective revenue security system to ensure all products sold generate sales revenue for the property.

Find Out More

Unscrupulous foodservice employees can steal from their employers, but they can also steal from an operation's guests.

In many cases, foodservice customers paying their bills hand over their credit cards to their servers, but they may not see where their card goes. This leaves opportunities for a guest's credit card information to be stolen on a credit-card skimming device that reads the magnetic stripe on a credit or debit card. When inserted into a card reader by a dishonest employee, the employee stores the card number, expiration date, and cardholder's name for later illegal use by the dishonest employee. Alternatively, an employee may also use a cell phone to photograph the front and back of a guest's credit card to obtain the name, credit card number, and the 3- or 4-digit security code number found on the back of the card.

Foodservice operators have a responsibility to help ensure the security of their guest's payments. Operators should stay updated about methods that could defraud their guests. To learn more about what can be done to help ensure the security of each guest's payment information, enter "preventing credit card theft by restaurant employees" in your favorite search engine and view the results.

What Would You Do? 11.1

"The beauty of our system," said Tim Liston, "is that you can monitor the actions of all your employees and supervisors."

Tim was talking to Gene Morrison, the owner of Fazziano's Italian Kitchen. Gene had called POS-Video Security, the company Tim represented, because, for the second time this year, Gene had discovered a case of employee/supervisor collusion. Working together, the employee and supervisor stole revenue from their restaurant by manipulating their unit's POS system.

"I'm pretty sure I understand your system, but go over it one more time," said Gene.

"Okay, Gene," said Tim. "Essentially, our new system goes beyond traditional surveillance methods by synchronizing the video being recorded with the data mined from your POS system to create detailed, customized video reports. Potentially fraudulent activity such as manager overrides, coupons or comps, or even a cash drawer being open for too long, is tracked, and the corresponding video surveillance can be searched by transaction number. Data reports and streaming video, both real time and stored, can be accessed securely and remotely on a PC, smart device, or in the cloud."

"So, for example," replied Gene, "When a sales void occurs, your system identifies the portion of video that was recording at the time of the void and then allows me to view just that portion of the video so I could see what was happening in the restaurant during the transaction."

"Exactly," said Tim.

Assume you were Gene. What types of employee fraud do you think could be uncovered using the technology offered by POS-Video Security's new product? Do you think the behavior of most dishonest cashiers and supervisors would change if they knew their actions were being video recorded? Explain your answer.

Developing a Revenue Security Program

Chapter 3 of this book introduced the basic profit formula:

$$\text{Revenue} - \text{Expense} = \text{Profit}$$

A close look at the formula might lead some foodservice operators to think that 50% of their time should address managing and protecting revenue, and 50% of their time should consider managing expenses. This may be reasonable because all cost control systems will be of little use if they cannot initially collect the revenue their businesses generates, deposit that revenue into their bank accounts, and spend it only on legitimate expenses.

Errors in revenue collection or other asset security issues can result from simple employee mistakes or, sometimes, outright theft by guests or employees. An important part of every operator's job is to devise revenue control systems to protect income. This is true regardless of whether it is cash, checks, credit or debit card receipts, coupons, meal cards, or another guest payment method.

Objectives of Internal Revenue Control

The American Institute of Certified Public Accountants (AICPA) has a longstanding definition of internal control:

> *Internal control comprises the plan of organization and all of the coordinate methods and measures adopted within a business to safeguard its assets, check the accuracy and reliability of its accounting data, promote operational efficiency, and encourage adherence to prescribed managerial policies.*[1]

This definition indicates that internal control relies on an organization plan and use of methods and measures to attain four key objectives:

1) Safeguard assets.
2) Check accuracy/reliability of accounting data.
3) Promote operational efficiency.
4) Encourage adherence to prescribed managerial policies.

Safeguard Assets
This objective addresses the protection of assets from losses including theft, resource maintenance (especially equipment) to ensure efficient utilization, and

---

1 American Institute of Certified Public Accountants, Committee on Auditing Procedures, Internal Control: Elements of a Coordinated System and Its Importance to Management and the Independent Public Accountant (New York, 1949).

the safeguarding of resources (especially inventories) to prevent pilfering, waste, and spoilage.

Check Accuracy/Reliability of Accounting Data

This objective addresses the checks and balances within the accounting system designed to ensure the accuracy and reliability of information. Dependable data is needed for reports to owners, government agencies, and other outsiders, and it is needed for an operator's internal use. Foodservice operations of all sizes benefit from the use of the Uniform System of Accounts for Restaurants (USAR). Accounting information is most useful when it is timely so internal control reports for operators must be prepared regularly and promptly.

Promote Operational Efficiency

A foodservice operation's training program and procedures with effective supervision promote operational efficiency. For example, when cooks are properly trained and supervised, food costs are lower. Similarly, the use of mechanical and electronic equipment improves efficiency. For example, POS devices can relay orders from server stations to preparation areas and increase staff efficiency while recording all guest purchases.

Encourage Adherence to Prescribed Managerial Policies

Asset security procedures should be designed to encourage compliance. For example, most foodservice operations have a policy that requires hourly paid employees to clock in and out personally; an employee cannot do this for another employee. Locating the time clock for easy managerial observation may encourage workers to follow the policy.

These four objectives may seem to conflict at times. For example, procedures for safeguarding an operation's assets may be so detailed that efficiency is reduced. Requiring multiple signatures to withdraw products from inventory may reduce theft, but the policy may require additional time and increased labor costs that exceed potential inventory losses.

Perfect controls, even if possible, may not be cost-justified. Foodservice operators must evaluate the cost of implementing procedures against their benefits. The reason: An operation's internal control procedures are most efficient when costs and benefits are balanced.

Elements of Internal Revenue Control Systems

Regardless of the specific foodservice operation, all effective internal control systems have similar elements:

Proper Leadership

Leadership is critical to a foodservice operation's internal control system. Effective policies must be developed, clearly communicated, and consistently enforced. Each level of management must be responsible for ensuring that applicable control procedures are adequate, and that any exceptions to policies should be minimized and justified.

Assigned Responsibility

When possible, responsibilities for a specific control activity should be given to a single individual. Then the staff member can be given a set of standard operating procedures with the expectation they will be followed. When responsibility is given to one person, an operator knows where to start looking if a problem is identified.

For example, a cashier should be solely and fully responsible for a specific cash bank. To ensure this, no one except that employee should have access to the bank during the cashier's shift. There should be no sharing of the bank and/or responsibility for it.

Separation of Duties

Separation of duties occurs when different personnel are assigned to accounting, asset responsibility, and production activities. Duties within the accounting function should also be separated. For example, different personnel should maintain an operation's general ledger (see Chapter 2) and the operation's **cash receipts journal**.

Key Term

Cash receipts journal: An accounting record used to summarize transactions related to cash receipts generated by a foodservice operation.

The major objective of separating duties is to prevent and detect errors and theft. Unfortunately, it is not always possible to separate duties in very small foodservice operations. In these situations, management must assume multiple duties.

Approval Procedures

Foodservice operators must authorize every business transaction. Authorization may be either general or specific. General authorization occurs when all employees must comply with selected procedures and policies when performing their jobs.

For example, in all operations servers must sell food and beverage products at the prices listed on the menu. Specific variations from this general policy, and any other significant variations, must be approved by management.

Proper Record Keeping

The accurate recording of security-related information is essential for effective internal control. Asset security documents including purchase orders, inventory evaluation sheets, sales records, and payroll schedules should be designed to be easy to complete and understand.

Written Policies and Procedures

Employees can only be expected to follow policies they understand. Significant security-related policies and procedures should be put in writing and be included in employee manuals and new employee orientation programs for all employees. Written policy and procedure manuals also help ensure all employees are treated fairly and similarly if policy violations occur.

Physical Controls

Physical controls are often necessary to properly safeguard assets. Examples include security devices such as cash safes and locked storage areas.

Performance Checks

Performance checks help to ensure all elements in the internal control system are functioning properly. Whenever possible, the checks should be independent (i.e., the person doing an internal verification must not be the same person responsible for collecting the data initially).

In large operations, auditors who are independent of both operations and accounting personnel report directly to top management. In smaller operations, managers should conduct most, but not all, performance checks. For example, an operation's **bank reconciliation** should be done by personnel who are independent from those who initially account for cash receipts and vendor payments.

Key Term

Bank reconciliation: A process performed to ensure that a company's financial records are correct. This is done by comparing the company's internally recorded amounts with the amounts shown on its bank statement. Any differences must be justified. When there are no unexplained differences, accountants can affirm that the bank statement has been reconciled.

Implementing and Monitoring a Revenue Security Program

In its simplest form, revenue control and security involve matching products sold with money received. Implementing and monitoring a total revenue security program involves ensuring that systems and recordkeeping are in place to allow foodservice operators to always verify that:

1) Documented Product (Menu Item) Requests = Product (Menu Item) Issues
2) Product Issues = Total Guest Charges
3) Total Guest Charges = Sales Receipts
4) Sales Receipts = Sales (bank) Deposits
5) Sales (bank) Deposits = Funds Available to Pay Legitimate Accounts Payable (AP)

To illustrate these five verification points required to ensure an effective revenue security program, consider Ferazia, who operates a Lebanese restaurant in New York City.

Ferazia considers her restaurant to be a family-oriented establishment. It has a small 20-seat cocktail area and 100 guest seats in the dining room. Total revenue at Ferazia's exceeds $1 million per year. When she started the restaurant, she did not give much thought to the design of her revenue control systems because she was generally in the restaurant. Due to her success, she now spends more time developing a second restaurant and needs both the security of an adequate revenue security system and the ability to review it quickly.

Ferazia has begun to develop a revenue security system, concentrating her efforts on ensuring that:

Product (Menu Item) Issues = Total Guest Charges = Sales Receipts = Sales (bank) Deposits = Funds Available to Pay Legitimate Accounts Payable (AP)

Ferazia knows that the first goal she must achieve is that of verifying her product issues.

Verification of Product Issues

A key to verification of product issues in the revenue security system is to follow one basic rule: *No menu item shall be issued from the kitchen or bar unless a permanent record of the issue is made.*

This means that the kitchen (or bar) should not fill any server request for menu items unless that product request has been documented in writing or electronically. In some small restaurants, the server's request for food or beverages takes the form of a single (or multicopy) written guest check, designed specifically for the purpose of revenue control. The top copy of this multicopy form generally is sent to the kitchen or bar. The guest check, in this case, becomes the documented request for the food or beverage product.

This "paper-only" system can work, but it is subject to some forms of abuse and fraud. Therefore, foodservice operations should utilize a POS system in which the "guest check" consists of an electronic record of product requests and issues. Then, a guest's order is viewed by the production staff on a computer screen or, in other cases, the POS system prints a hard copy of the order used by the production staff.

In either case, the software within the POS system creates a permanent record of the transaction and issues a unique transaction number to identify a requested product. This record authorizes the kitchen to prepare food or the bar to make a drink. If a foodservice operation elects to supply its employees with meals during work shifts, these meals should also be recorded in the POS system.

In the bar, the principle of verifying all product sales is even more important. Bartenders should be instructed to never prepare a drink unless that drink has first been recorded in the POS system. This should be the procedure, even if the bartender is working alone.

This rule regarding product issuing is important for two reasons. First, requiring a permanent documented order ensures there is a record of each product's sale. Second, this record of product sales can be used to verify both proper inventory usage and product sales totals.

Ferazia enforces this basic rule by requiring that no item be served from her kitchen or the bar without the sale first being entered into her POS system. If her verification of product sales system works correctly, Ferazia will note the following formula is always correct:

Documented Product Requests = Product Issues

Technology at Work

A high-quality POS system always provides detailed sales reports. The sales report dashboard on a POS system provides an overview of all transactions completed during a selected time. This includes net sales, service charges, tips, total guests served, table turn times, and a breakdown of all service types and payment methods. It should also provide detailed sales exception reports that allow users to quickly see an overview of all voids, discounts, and refunds. The reports also allow users to identify the specific servers and managers who are giving and approving sales exceptions—voiding receipts, discounting food, and offering refunds.

It is important to recognize that, regardless of its level of sophistication, a POS system will not "bring" control to a foodservice operation. A high-quality POS system can, however, take good control systems that have been carefully designed by management and add to them speed, accuracy, and/or additional information.

Properly utilized, a POS system is of immense value. If, however, an operation has no formal revenue security plan, the POS system simply becomes a high-tech adding machine used primarily to sum guest purchases and nothing more. Properly selected and utilized, however, POS systems play a crucial role in the implementation of an operation's complete revenue security system.

To review the cash control-related features of some currently popular POS systems, enter "POS revenue control reports for restaurants" in your favorite search engine and view the results.

Experienced foodservice operators know that despite their best efforts, it is possible for employees to issue menu items without a documented product request when:

1) Two or more employees work together to defraud the operation. Collusion of this type can be discovered when operators use a system to carefully count the number of items removed from inventory and then compare that number to the number of products actually issued.

2) A single employee (such as a bartender working alone) is responsible for both making and filling the product request. Operators can uncover this fraud when they carefully compare the number of items (or beverage servings) removed from inventory with the number of recorded product issues.

Verification of Guest Charges

When the production staff is required to prepare and distribute products only in response to a properly documented request, it is critical that the documented requests result in charges to the guest. The reason: It makes little sense to enforce verification of product issues without also requiring the service staff to ensure that guest charges match these requests.

There are several ways to achieve this requirement. If an operator insists that no product be issued without a POS-generated request, the managerial goal is to ensure that product issues equal guest check totals. In other words, all issued products should result in appropriate charges to the guest.

When properly implemented, this second step of the revenue control system will ensure the following formula always holds true:

Product Issues = Total Guest Charges

Ferazia has now implemented the first two key revenue control principles. The first one is that no products can be issued from the kitchen or bar unless the order is documented; the second one is that all guest charges must match product issues.

With these two systems in place, Ferazia can deal with many problems. If, for example, a guest has "walked" their check, the operation has a duplicate record of the transaction. The POS system would have recorded which products were sold to this guest, which server sold them, and perhaps additional information such as the time of the sale, the number of guests in the party, and the sales value of the products.

The POS system Ferazia uses also ensures that service personnel cannot "change" the prices charged for items sold. Note: This would likely be possible in an operation using manual (paper) guest checks.

To complete this aspect of her control system, Ferazia also implements a strict policy regarding the documentation of employee meals. Doing so has the added advantage of providing a monthly total of the value of employee meals. Recall that this amount is needed to accurately compute cost of food sold (see Chapter 3) when preparing the property's income statement.

Ferazia is now ready to address the next major component in her revenue security system. That component is the actual collection of guest payments. These payments will represent Ferazia's actual sales receipts.

Verification of Sales Receipts

Sales receipts refer to the actual revenue received by the cashier, server, bartender, or other designated personnel in payment for products served. In Ferazia's case, this means all sales revenue from her restaurant and lounge.

The essential principle to recognize in this step is that it requires two individuals (the cashier and a member of Ferazia's management team) to verify sales receipts. Although this will not prevent possible collusion by a pair of individuals, it is important that sales receipt verification is a two-person process. Ferazia wants to ensure that the amount of cash collected, when added to her non-cash (credit and debit card) guest payments matches the dollar amount she has charged her guests as shown in the POS system.

In most operations, individual guest charges are recorded only in the POS system. This is the case, for example, in a QSR or cafeteria, where food purchases are totaled and paid for at the same time. In these instances, the POS system provides an accurate total of guest and other charges. Receipts collected should always equal these charges. If Ferazia's revenue security system is working properly, the following formula noted above will always be true:

Total Guest Charges = Sales Receipts

Note that total POS recorded charges consist of all sales, service charges, tips, and guest-paid taxes, which are the total revenue the operation should receive.

Verification of Sales Deposits

Most foodservice operations make a sales deposit each day the property is open because keeping excessive amounts of cash on hand is not advisable. It is strongly recommended that only management make the actual bank deposit of daily sales receipts. A cashier or other clerical assistant may complete a deposit slip, but a manager should be responsible for monitoring the deposit of sales.

This task involves the actual verification of the contents of the deposit and the process of matching bank deposits with actual sales receipts. These two numbers should match. That is, if Ferazia, or a member of her management team, deposits Thursday's sales on Friday, the Friday deposit should match the sales amount of Thursday. If it does not, her operation has experienced some loss of revenue.

It is this step of the revenue control system in which embezzlement is most likely to occur. Embezzlement is a crime that often goes undetected for long periods of time because many times the embezzler is a trusted employee. Falsification of, or destroying, bank deposits is a common method of embezzlement.

To prevent this activity, Ferazia should take the following steps to protect her sales deposits:

✓ Make bank deposits of cash and checks daily, if possible.

✓ Ensure the person preparing and making the deposit is bonded.

✓ Establish written policies for completing bank reconciliations: the regularly scheduled comparison of the business's deposit records with the bank's deposit records.

✓ Payment card funds transfers to a business's bank account should be reconciled each time they occur. Today, in most cases, cash and non-cash payment reconciliations can be accomplished daily using online banking features.

✓ Review and approve written bank statement reconciliations at least monthly.

✓ Change combinations on office safes periodically and share the combinations with the fewest possible employees.

✓ Require all cash handling employees to take regular and uninterrupted vacations at least annually so another employee can assume and uncover any improper practices.

✓ Consider employing an outside auditor to examine the accuracy of deposits on an annual basis.

If the verification of sales deposits is done correctly, and no embezzlement is occurring, the following formula should always hold true:

$$\text{Sales Receipts} = \text{Sales (Bank) Deposits}$$

Sales deposit records are maintained in an operation's **back office accounting system,** and these must be regularly and carefully monitored by an operation's managers or owners.

Key Term

Back office accounting system: The accounting software used to maintain a business's accounting records that are not contained in its POS system. This typically includes items such as payroll records, accounts payable and receivable, taxes due and payable, and net profit and loss summaries, as well as all balance sheet entries.

Technology at Work

Modern POS systems are an important tool in maintaining revenue control, but they are not the only important tool. To ensure an effective revenue control system foodservice operators must also have an up-to-date and effective back office accounting system.

> The term "back office" originated when early companies designed their offices so that the front portion contained the employees who interact with guests. The back portion of the office contained associates with little interaction with guests such as accounting clerks.
>
> Back office accounting software is used by foodservice businesses to help track income and expenses, to create invoices, calculate sales tax, price recipes and menus, and more.
>
> The best back office accounting systems can be integrated (electronically connected) to an operation's POS, making it easy to monitor both revenue collection and revenue deposits and cash account maintenance.
>
> To learn more about the features included in these essential accounting systems, enter "back office accounting systems for restaurants" in your favorite search engine and review the results.

Verification of Accounts Payable (AP)

AP as defined in this step refers to the legitimate amount owed to a vendor for the purchase of products or services. The basic principle to be followed when verifying AP is: *The individual authorizing the purchase should verify the legitimacy of the vendor's invoice before it is paid.*

Vendor payments are often an overlooked potential threat to the security of a foodservice operation's revenue. Of course, an operation should pay all valid expenses. However, both external vendors and an operation's employees can attempt to defraud a foodservice operation by invoices.

For example, consider again the case of Ferazia. She has just received an invoice for fluorescent light bulbs. The invoice is for over $400 dollars, yet the invoice lists only two dozen bulbs were delivered (a large overcharge).

Ferazia is not familiar with this specific vendor, but the delivery slip included with the invoice was signed (six weeks ago!) by her receiving clerk. Quite likely, in this case, Ferazia and her operation are the victims of an invoice scam by the vendor, and the scam threatens her operation's revenue.

Find Out More

Foodservice operations are popular targets for invoice fraud. In these scams, criminals send bills for goods or services a business never ordered or received. The scam succeeds mainly because the invoices look legitimate, and unsuspecting accounts payable employees do not look closely to determine if they are real. They simply make the payment, thinking someone else in the operation placed the order.

(Continued)

To protect your operation from invoice scams, you must:

1) Be cautious when processing invoices: Ensure your accounts payable personnel are aware of these scams and are cautious when processing invoices.
2) Verify unfamiliar vendors: Do not purchase from new suppliers or pay invoices from unfamiliar vendors until you verify their existence and reliability.
3) Check invoices against original purchase orders (POs) before paying them: A written PO (see Chapter 8) should exist for each invoice an operator receives.

To learn more about additional procedures and policies operators can implement to minimize their chances of becoming a victim of invoice fraud, enter "protecting against invoice fraud" in your favorite search engine and review the results.

Dishonest suppliers can take advantage of weaknesses in an organization's purchasing procedures or of unsuspecting employees who may not be aware of their fraudulent practices. In addition, the supplies delivered by these bogus firms are most often highly overpriced and of poor quality.

When her revenue security program is working properly, Ferazia can confirm that:

Sales (bank) Deposits = Funds Available to Pay Legitimate Accounts Payable (AP)

Funds available for AP should only be used to pay legitimate expenses that result from a purchase verified by authorized personnel within the foodservice operation.

When Ferazia properly completes the building of her revenue control system, its key features can be summarized as shown in Figure 11.3.

---

1) No product shall be issued from the kitchen or bar unless a permanent record of the issue is made.
2) Product issues must equal total guest charges.
3) Both the cashier and a member of management must verify all sales receipts.
4) Management must personally verify all bank deposits.
5) Management or the individual authorizing the purchase should verify the legitimacy of all vendor invoices before they are paid.

Figure 11.3 Revenue Control Points Summary

The best revenue control programs primarily help foodservice operators maintain the security of their cash. In most cases, however, the largest percentage of a foodservice operation's assets is represented by its investment in furniture, equipment, building, and land. These resources are expected to benefit a foodservice operation for at least one year and are referred to as fixed assets (see Chapter 3).

Fixed assets are managed differently from current assets whose benefits will be realized in less than one year. The next chapter will address those differences, and how foodservice operators can choose accounting professionals to help when accounting support is needed or desired.

What Would You Do? 11.2

Dick Wright worked for 15 years as the head snack bar cashier for the Downtown Sports Arena Complex, a facility whose food concessions were managed by Susan Harper's "Elite Catering" company.

Dick had twice won the company's "Employee of the Year" award, and Susan considered Dick to be a valued and trusted employee who had, on many occasions, performed above and beyond the call of duty.

Susan was very surprised when newly installed video surveillance equipment confirmed that Dick, despite strict written rules against it, had, on several recent occasions, given free food and beverages to friends of his who visited the arena.

When confronted with the video evidence, Dick admitted the conduct, apologized profusely, and asked Susan for a second chance. He promised never to give free food or beverages to anyone in the future.

On the advice of the company's attorney, Susan is documenting, in writing, her decision on handling the situation.

Assume you were Susan. Do you believe an employee caught defrauding their employer should ever be given a second chance? If so, under what circumstances? What impact will your decision in this case have if, in the future, other employees are caught stealing from your company?

Key Terms

Fraud
Embezzlement
Walk (bill)
Open check

Short (cash bank)
Over (cash bank)
Bonding (employee)
Cash receipts journal

Bank reconciliation
Back office accounting
   system

Operator's 10-Point Tactics for Success Checklist

Evaluate your need for, and the current status of, each of the following operational tactics. For those tactics you think are important, but not yet in place, develop an action plan for its implementation including who will be responsible for the tactic's completion and the target date by which it should be completed.

				If Not Done	
Tactic	Don't Agree (Not Done)	Agree (Done)	Agree (Not Done)	Who Is Responsible?	Target Completion Date
1) Operator understands the importance to profits of developing an effective revenue control program.	——	——	——		
2) Operator can identify external threats to revenue security in their own operation.	——	——	——		
3) Operator can identify internal threats to revenue security in their own operation.	——	——	——		
4) Operator can state the objectives of an internal revenue security program.	——	——	——		
5) Operator can summarize the elements that must be present in an effective revenue control system.	——	——	——		
6) Operator recognizes the importance of continually monitoring and verifying their product issues.	——	——	——		
7) Operator recognizes the importance of continually monitoring and verifying their guest charges.	——	——	——		
8) Operator recognizes the importance of continually monitoring and verifying their sales receipts.	——	——	——		

				If Not Done	
Tactic	**Don't Agree (Not Done)**	**Agree (Done)**	**Agree (Not Done)**	**Who Is Responsible?**	**Target Completion Date**
9) Operator recognizes the importance of continually monitoring and verifying their sales deposits.	——	——	——		
10) Operator recognizes the importance of continually monitoring and verifying their accounts payable (AP).	——	——	——		

12

Accounting for Fixed and Other Assets

What You Will Learn

1) How to Account for Fixed and Other Assets
2) How to Depreciate and Exchange Fixed Assets
3) How to Choose Professional Accounting Assistance

Operator's Brief

The largest percentage of a foodservice operation's assets is usually represented by its investment in furniture, equipment, building, and land. These are called fixed assets and are managed differently from current assets.

Foodservice operators often have the choice of purchasing or leasing fixed assets. The decision to buy or lease affects how an asset is accounted for, and this is carefully explained in this chapter.

Depreciation is a key concept that must be understood when accounting for many types of fixed assets. In this chapter, you will learn about the two most-used approaches to assessing depreciation costs and the procedures for computing depreciation values using each approach.

When a foodservice operator decides to dispose of a fixed asset, regardless of its value, accounting procedures used to record this disposal relate to the depreciation methods previously used. In this chapter, the disposal of fixed assets is thoroughly reviewed as are the procedures for recording the asset's sale (or, if appropriate, scrapping the equipment item).

Sometimes, a fixed asset is traded in for a newer or improved asset, and this chapter explains how to properly account for this asset exchange.

Lastly, accounting for a foodservice operation can be a very time- and knowledge-intensive process. Therefore, the chapter will explain helpful procedures if you decide to seek outside professional help to manage the accounting needs of your operation and provides some insight into future issues of importance to managerial accountants working in the foodservice industry.

CHAPTER OUTLINE

Accounting for Fixed Assets
Recording Fixed Asset Purchases
Depreciating Fixed Assets
    Straight-line Depreciation
    Double Declining Balance Depreciation
Other Issues Related to Fixed Assets
    Uniforms, Linens, China, Glass, Silver, and Utensils
    Disposal of Fixed Assets
    Exchange of Fixed Assets
Accounting for Other Assets
Choosing Professional Accounting Assistance
Monitoring Evolving Accounting Issues

Accounting for Fixed Assets

Fixed assets include those that are depreciated (buildings and equipment) and others that are not depreciated (land). Fixed assets typically constitute the largest percentage of the total assets of a foodservice operation. The Uniform System of Accounts for Restaurants (USAR) uses the following classifications for fixed assets:

✓ Land
✓ Buildings
✓ Costs of improvements in progress
✓ Leaseholds and leasehold improvements
✓ Furniture, fixtures, and equipment (FF&E)
✓ Uniforms
✓ Linens
✓ China
✓ Glass
✓ Silver
✓ Utensils

Expenditures for fixed assets are **capital expenditures** rather than **revenue expenditures**.

Revenue expenditures are those made by a foodservice operation in which the benefits are expected to be received within a year or less. Examples include wage payments and utility expenses.

In contrast, capital expenditures recognize that, when a foodservice operation is built, its

Key Term

Capital expenditure:
Expenditures made by a foodservice operation in which benefits are expected to be received for a period greater than one year.

Key Term

Revenue expenditure:
Expenditures made by a foodservice operation in which benefits are expected to be received within one year.

owners generally expect to realize financial benefits over many years. Capital expenditures, then, are expenditures for which benefits are expected to be received over a period greater than one year. Capital expenditures must not be recorded as revenue expenditures because of the impact capital-related entries have on the income statement (see Chapter 3).

If a capital expenditure with a life of 10 years is improperly recorded as a revenue expenditure, all the expense will be recorded in one year instead of being spread over 10 years. This would distort the foodservice operation's income statement for all 10 years because net income for year one would be understated, and net income for years two through 10 would be overstated.

Occasionally, a foodservice operation may purchase an item for a small amount that is technically a capital expenditure but, instead, it is treated as a revenue expenditure. An example is the purchase of a kitchen knife that costs $25.00. The knife may last many years, but the time required to calculate annual depreciation on it is not worth the effort to spread this minimal cost over, for example, 10 years.

The purchase of fixed assets should be recorded at their actual cost including all reasonable and necessary expenditures required to properly make the asset operable. Examples of these expenditures are items including freight charges, sales tax, and installation charges. Charges not considered reasonable and necessary might include removal of minor dents in metal equipment purchased in used condition, and traffic tickets incurred by the delivery driver.

Recording Fixed Asset Purchases

As an example of how fixed asset purchases are recorded, assume that Maureen, the owner of Maureen's Irish Pub, purchased kitchen equipment at a list price of $50,000 with terms of 2/10, n/30 (2% discount if paid within 10 days; net amount due in 30 days). Maureen has enough cash to pay within the cash discount period. Also assume that sales tax of $2,000, freight charges of $600, and installation costs of $500 must be paid to put the equipment in operation. Maureen also purchases a two-year maintenance contract for a total cost of $2,000.

The cost to be recorded in this example would be:

Equipment list price	$50,000
Less 2% cash discount	(1,000)
Net price	$49,000
Sales tax	2,000
Freight charges	600
Installation charges	500
Total cost of equipment	$52,100

In this example, the maintenance contract is *not* added to the equipment cost, and it is not depreciated over the asset's life. Instead, the $1,000 for the next 12 months is recorded by Maureen as a prepaid expense, and the $1,000 payment for the second year is recorded as deferred insurance (another type of asset).

When land is purchased, all reasonable/necessary expenditures to buy the land are included in the purchase price. These include property taxes paid, title opinions, surveying costs, brokerage commissions, and any required excavating expenses.

Sometimes, a foodservice operation leases an asset under an agreement requiring it to record the lease as a **capital lease**. Then, the leased asset is recorded as an asset on the lessee's books and is included under furniture, fixtures, and equipment (FF&E) as "Leased asset under a capital lease."

When a foodservice operator leases a building, they must often make extensive improvements to walls, ceilings, lighting, and similar interior items. These are called leasehold improvements (see Chapter 4). The cost of the leasehold improvements is amortized over the remaining life of the lease, or the life of the improvement, whichever is shorter.

Key Term

Capital lease: A lease agreement of long duration, generally noncancelable, in which the lessee (foodservice operation) assumes responsibility for property taxes, insurance, and maintenance of the leased property.

To illustrate, assume that an operator leases a building for 10 years and spends $50,000 on improvements that have a five-year life. The operation would amortize $10,000 of the cost of the improvements each year for five years. The debit (left side) and credit (right side) entries to the operation's general ledger would be:

Amortization of leasehold improvements	$10,000	
Leasehold improvements		$10,000

The "Cost of improvements in progress" is another fixed asset reported on an operation's balance sheet. This represents all labor, materials, advances on contracts, and interest on construction loans incurred in the current construction of property/equipment. "Uniforms, linens, china, glass, silver, and utensils" is a unique fixed asset account found on the balance sheet of many operations. This asset category is important to control. If these items are carelessly discarded or stolen, for example, expenses will increase, and net income will decrease. (Note: This category of fixed assets will be addressed in greater detail later in this chapter.)

Depreciating Fixed Assets

In Chapter 2, depreciation was defined as "the allocation of the cost of equipment and other depreciable assets based on the projected length of their useful life." To restate, depreciation is the reduction in the net book value of a fixed asset.

The total amount of depreciation that can be taken equals the cost of the fixed asset minus its scrap or salvage value (see Chapter 2). Since a foodservice operator does not write a check for depreciation, it is different from most other expenses because it is a non-cash expense.

Depreciation is not a fund of cash set aside by an operation's owner, but it does save cash for the owner because it provides a tax shelter by reducing taxable income. Property and equipment values are typically shown on the books at their net book value (see Chapter 4). This is determined when the accumulated depreciation to date is subtracted from the cost of the asset. Recall that accumulated depreciation is a contra asset account, and it always carries a credit balance (see Chapter 2).

Two methods of depreciation are typically used to account for fixed assets in the foodservice industry:

✓ Straight-line depreciation
✓ Double declining balance depreciation

Straight-Line Depreciation

Straight-line depreciation is a simple form of calculating depreciation for a fixed asset. It simply requires an operator to divide the cost of the asset by the estimated number of years of its life.

When utilizing the straight-line depreciation method to calculate the annual depreciation, an operator takes the cost of the asset less its salvage value (see Chapter 2) and divides the remaining amount by the estimated life of the asset.

Key Term

Straight-line depreciation: A method of depreciation that distributes the expense evenly over the estimated life of a fixed asset.

For example, assume that dining room furniture costs an operator $34,000, has a salvage value of $4,000 and an estimated life of five years. The annual depreciation under the straight-line method of depreciation would be $6,000 per year:

Cost of FF&E	$34,000	
Less: Salvage value	(4,000)	
Depreciable cost	$30,000	÷ 5 years = $6,000 annual depreciation

Total depreciation over life of asset: $6,000 annual depreciation × 5 years = $30,000

Double Declining Balance Depreciation

Double declining balance depreciation (DDB) is an **accelerated method of depreciation**.

When using the DDB method, the result is a higher depreciation charge in the first year with lower and lower charges in successive years. The DDB method does not take salvage value into account. While the DDB method is more complicated to calculate, it has the benefit of higher initial levels of depreciation with reduced taxable income levels during the early years of the equipment's life.

To use the DDB method, operators utilize three steps:

Step 1: Determine the straight-line (SL) depreciation rate.

$$\text{Straight-line rate} = \frac{1}{\text{Number of years}}$$

Step 2: Multiply the SL depreciation rate by 2 to determine the DDB percentage.

$$\text{SL rate} \times 2 = \text{DDB}\%$$

Step 3: Multiply the undepreciated amount of the fixed asset by the DDB% to determine the annual depreciation expense.

$$\text{DDB}\% \times \text{Undepreciated amount} = \text{Depreciation expense}$$

Key Term

Double declining balance depreciation (DDB): An accelerated method of depreciation that ignores the salvage value of a fixed asset and yields a greater amount of depreciation in the early years of the asset's life.

Key Term

Accelerated method of depreciation: A method of depreciation that results in higher depreciation charges in the first year; charges gradually decline over the life of a fixed asset.

To illustrate the differences between the straight-line and DDB depreciation methods, consider the case of Nancy Graves. For her soon-to-open ghost restaurant, Nancy has just purchased a new POS terminal for $4,000. The estimated useful life of the terminal is 8 years with a salvage value of $400.

Nancy's annual depreciation for years one through eight for the two depreciation methods is shown in Figure 12.1. Note that the DDB yields Nancy the greatest amount of annual depreciation for years one through three, while the straight-line method results in the greatest amount of depreciation for each of the last five years.

$4,000 Fixed Asset Depreciation

Year	Straight-Line	DDB
1	($4,000 – $400) ÷ 8 = $450	1/8 = 12.5%; 12.5 × 2 = 25% $4,000 × 0.25 = $1,000
2	$450	($4,000 – $1,000) × 0.25 = $750
3	$450	($3,000 – $750) × 0.25 = $562.50
4	$450	(2,250 – $562.50) × 0.25 = $421.88
5	$450	($1,687.50 – $421.88) × 0.25 = $316.40
6	$450	($1,265.62 – $316.40) × 0.25 = $237.30
7	$450	($949.22 – $237.30) × 0.25 = $177.98
8	$450	($711.92 – $177.98) × 0.25 = $133.48
Total Depreciation	$3,600	$3,599.54

Figure 12.1 Annual Depreciation with Straight-line and Double Declining Balance Methods

Both the straight-line and DB methods can be used to depreciate fixed assets. The method selected by a foodservice operator should be the one that best reflects the decline in value of the fixed asset being depreciated.

Technology at Work

Maintaining up-to-date depreciation values for fixed assets is important, but it can be complicated and time-consuming. To assist in this process, inexpensive computer programs have been developed that conform to all generally accepted accounting principles (GAAP).

These fixed asset software programs automate the process of tracking fixed assets through the various stages of the asset life cycle from acquisition through depreciation and then their disposal.

To view some features and prices of these helpful programs, enter "fixed asset software for restaurants" in your favorite search engine and view the results.

The generally accepted accounting principle (GAAP) of *consistency* (see Chapter 1) requires that the method selected for use with a specific fixed asset should continue to be used over the life of that asset. It is, however, acceptable to use different depreciation methods for different fixed assets. For example, a building may be depreciated with the straight-line method, and the DDB method could be used for one or more kitchen equipment items.

What Would You Do? 12.1

"I don't understand," said Casey, "All our sales projections and costs were in line with the budget. How could we miss our profit goals?"

Casey, the kitchen manager at Buffalo Bob's Bistro, was talking with Andre, the restaurant's manager. They had just received the operation's income statement for March, and they missed their profit forecast.

"Well," replied Andre, "it has to do with the new refrigerators and freezers we installed last month. The owners decided to depreciate those pieces of equipment using an accelerated depreciation method. That's why our depreciation expense is so high, and why our profits are lower than we forecasted."

"O.K., but I still don't see how their decision about depreciation can make us lose money," said Casey.

Assume you were one of the owners of Buffalo Bob's Bistro. What would be your rationale for using a depreciation method yielding lower profit in the earlier years and then smaller losses in the later years of equipment lives?

Other Issues Related to Fixed Assets

Foodservice operators must address other concerns regarding accounting for their fixed assets. Among the most important of these are issues related to:

✓ Uniforms, Linens, China, Glass, Silver, and Utensils
✓ Disposal of Fixed assets
✓ Exchange of Fixed assets

Uniforms, Linens, China, Glass, Silver, and Utensils

Foodservice operators can use several methods to charge uniforms, linens, china, glass, silver, and utensils expenses to their operations including:

1) Considering these items part of fixed assets and physically counting and reflecting the total cost of the items on hand.
2) Capitalizing the initial stock of these items and then expensing the cost of the items as they are later brought out and placed into service.

Both methods have merit; however, for uniformity one method is preferable. Capitalize the cost of the initial stock of these items. Then, depreciate this capitalized cost over a period not to exceed 36 months and expense replacements when they are placed into service.

Reserve stocks of these items should be considered inventory until they are placed into service. The total accumulated depreciation should appear as a separate line item on the balance sheet. This amount is then subtracted from the total fixed asset line to determine the net book value of the business's fixed assets.

To illustrate, assume a foodservice operator purchased an initial stock of uniforms, linens, china, glass, silver, and utensils for $45,000 cash on November 30, 20XX, and they decide to write off the stock over 30 months. On December 15, replacement items were purchased for $3,000 but not placed into service. The journal entries from November 15 until December 31, 20XX, would be as follows:

	Debit	Credit
Uniforms, linen, china, glass, silver, and utensils	$45,000	
Cash		$45,000
Inventory: Uniforms, linen, china, glass, silver, and utensils	$3,000	
Cash		$3,000
Depreciation expense: Uniforms, linen, china, glass, silver, and utensils	$1,500	
Accumulated depreciation expense: Uniforms, linen, china, glass, silver, and utensils		$1,500

On December 31, 20XX, the operator's income statement would include $1,500 of depreciation expense for uniforms, linen, china, glass, silver, and utensils. The operator's balance sheet would include the following:

Under Current Assets:		
Inventory: Uniforms, linen, china, glass, silver, and utensils		$3,000
Under Fixed Assets: Uniforms, linen, china, glass, silver, and utensils	$45,000	
Less: Accumulated depreciation	($1,500)	$43,500

Disposal of Fixed Assets

Sometimes, a foodservice operator removes fixed assets from its books because the assets are being scrapped, sold, or traded-in. If a fixed asset is scrapped, and it has been fully depreciated, both the asset and related accumulated depreciation must be removed from the books.

To illustrate, assume a dish machine with a cost of $80,000 and a salvage value of zero was fully depreciated with the following accounting entry:

Accumulated depreciation	$80,000	
Dish machine		$80,000

If the dish machine is scrapped before being fully depreciated, a different entry is made. Assume that the same dish machine costing $80,000 had been depreciated to the extent of $60,000. The journal entry would be:

Accumulated depreciation	$60,000	
Loss on disposal of asset	$20,000	
Dish machine		$80,000

In this example, the loss representing disposal of the dish machine would be closed into the income summary account at the end of the next fiscal period.

In some cases, fixed assets are disposed of with a sale resulting in a gain or loss. Assume the dish machine in this example had a net book value of $12,000 based on a cost of $80,000 and an accumulated depreciation balance of 68,000. If it is sold for $16,000, the journal entry would be:

Cash	$16,000	
Accumulated depreciation	$68,000	
Dish machine		$80,000
Gain on disposal of asset		$4,000

The gain of $4,000 is the difference between the selling price of $16,000 and the book value of $12,000.

Now assume the same facts concerning the $80,000 dish machine, except it is sold for $8,000. Since it has a net book value of $12,000, and is sold for $8,000, a loss of $4,000 would result. In this scenario, the journal entry would be:

Cash	$8,000	
Accumulated depreciation	$68,000	
Loss on disposal of asset	$4,000	
Dish machine		$80,000

Exchange of Fixed Assets

A fixed asset such as a delivery van might be exchanged or traded in for a similar and newer asset. To illustrate, assume that a $30,000 van is exchanged along with $32,000 cash for a new van with a list price of $33,000. At the time of the exchange, the old $30,000 van had accumulated depreciation of $28,000. It has a book value of $2,000.

Although the van's book value is $2,000, the operator is receiving only $1,000 on the trade in. There is a loss of $1,000. The journal entry to record the exchange would be:

Delivery van (new)	$33,000	
Accumulated depreciation	$28,000	
Loss on disposal of asset	$1,000	
Cash		$32,000
Delivery van (old)		$30,000

Now assume the operation is given $5,000 on the trade in and must pay only $28,000 for the new delivery van. The journal entry to record this exchange would be:

Delivery van (new)	$30,000	
Accumulated depreciation	$28,000	
Cash		$28,000
Delivery van (old)		$30,000

The **Financial Accounting Standards Board (FASB)** has stated that no gains should be recorded on exchanges. Instead, the gain is to be reflected in the value of the asset acquired.

Two important points should be noted about the above journal entries:

1) The generally accepted accounting principle of *conservatism* states that losses, but not gains, should be recorded.
2) Tax reporting rules differ from financial reporting rules. When reporting for tax purposes, neither gains nor losses are recorded on the exchange of similar assets.

Key Term

Financial Accounting Standards Board (FASB): An independent, private-sector, not-for-profit organization that establishes financial accounting and reporting standards for public and private companies and not-for-profit organizations that follow GAAP.

Find Out More

The Financial Accounting Standards Board (FASB) is recognized by the U.S. Securities and Exchange Commission as the designated accounting standard setter for public companies. FASB standards are recognized as authoritative by many other organizations, including state Boards of Accountancy and the American Institute of CPAs (AICPA).

The FASB develops and issues financial accounting standards uses with a transparent and inclusive process to promote financial reporting providing useful information to investors and others who use financial reports.

To learn more about the activities of this important organization, enter "Financial Accounting Standards Board (FASB)" in your favorite search engine and view the results.

Accounting for Other Assets

The "Other Assets" category of the balance sheet (see Chapter 4) is reserved for items that cannot be included in another grouping. The most typical (and those included in the USAR) are for:

✓ Amounts paid for goodwill

Goodwill (see Chapter 4) is the excess of the purchase price beyond an asset's approved or normal value. It exists when expected future earnings exceed the normal rate for the foodservice industry. Goodwill is only entered into an operation's books when the operation is purchased.

✓ Cost of bar license

In states restricting the number of licenses to sell alcoholic beverage, an existing license must often be purchased.

✓ Rental deposits

Deposits of cash or marketable investments may be required as security for a building being leased and to assure compliance with other rental agreement terms. Other examples of deposits include those for utility services such as water, gas, and electricity.

✓ Cash surrender value of life insurance.

This is insurance carried on the lives of officers, partners, or key employees when the foodservice operation is the policy's **beneficiary**. The policy's **cash surrender value** should be accrued as an asset reflecting the increase in the cash surrender value as the life insurance premium is paid.

✓ Deposit on franchise or royalty contract

Deposits for these purposes are stated separately in the "Other Assets" section of a foodservice operation's balance sheet.

Key Term

Beneficiary: A person or entity who is the recipient of funds or other property under a will, trust, or insurance policy.

Key Term

Cash surrender value: Money an insurance company pays to a policyholder if the policy is voluntarily terminated before maturity, or an insured event occurs.

Choosing Professional Accounting Assistance

While the accounting needs of each foodservice operation are unique to that operation, every foodservice owner or operator must determine whether in-house staff maintain their accounting records, or if they employ outside professional assistance in the preparation of their financial documents.

If the decision is made to utilize external professional accounting assistance, it is important to do so carefully, and a variety of important issues must be considered. These include the accountant's location, the division of workload, and the type of accounting software to be used. Additional issues include the accountant's salary and benefits, and the types of decisions they can make to improve the business and/or lower its taxes.

It will always be in a company's best interest to choose an experienced and capable person or organization to assist in managing the company's finances. If the right assistance is chosen, it can save the company time and money year-after-year. Key issues to consider when selecting professional accounting assistance include:

✓ Location
✓ Certification
✓ Experience
✓ Assignments
✓ Software utilized
✓ Fees

Location

Until recently, it was important that a foodservice accountant be located near the operation so financial data could be readily shared. Today, however, increasing numbers of operators share their financial information online using cloud-based technology. As a result, the physical location of those who provide accounting assistance is of less importance. Utilizing **cloud accounting software,** the owner or operator and the accountant can view identical real-time data at the same time regardless of their locations.

The decision about where accounting assistance should be located most often depends upon the preferences of a business's owner. If, for

Key Term

Cloud accounting software: Software like traditional, on-premises, or self-installed accounting software, except the accounting software is hosted on remote servers. Data is sent into "the cloud," where it is processed and returned to the user.

example, a business owner in New York communicates by email, phone calls, video-conferences, and/or secure cloud accounting software, the accountant could be thousands of miles away in California.

If, however, the owner prefers face-to-face contact and finds it useful to have the accountant accompany them to selected business meetings, the search for professional assistance will be limited to those located nearby, or those willing to travel to meet with the owner at selected times.

Wherever they are based, an operator's accountant must be well-versed in the local laws affecting the operation. Examples of state or local regulations include those related to health codes, tips, overtime and wage payments, and unemployment tax withholdings. While the physical location of an accountant may vary, their knowledge of local regulations directly affecting a business must not.

Certification

To ensure they select a qualified individual to assist with financial records, many foodservice operators choose a **Certified Public Accountant (CPA)**. A CPA has been granted certification by the Board of Accountancy for the state, district, or country in which the CPA does business.

CPA requirements set by each state board of accountancy include completing a program of study in accounting at a college or university and passing the Uniform CPA Exam. It is also necessary to obtain a specific amount of professional work experience in public accounting. The amount and type of experience varies according to licensing jurisdiction.

> **Key Term**
>
> **Certified Public Accountant (CPA):** An accounting professional who has earned a CPA license through a combination of education, experience, and examination.

It is possible to use accountants who are not certified whose professional fees are less expensive than those of a CPA but doing so might be unwise. Tasks such as bookkeeping, tax preparation, and general financial management might not require a CPA, but a business will almost certainly need one if it ever seeks a loan or is audited by a taxing authority.

Find Out More

There is no legal requirement that a foodservice operation employ a Certified Public Accountant (CPA) to prepare financial records. In some cases, however, banks and investors may require that an operation's financial records be prepared by a CPA to meet their own lending or investment requirements.

In most cases, CPAs not only have tax preparation and financial planning expertise but they can also provide a business with useful advice and helpful

---

(Continued)

insights related to financial data collection and analysis. The best CPAs can also provide an additional resource to help business owners make the best long-term strategic planning decisions.

To learn more about the advantages of utilizing the services of a CPA in the preparation of a foodservice operation's financial documents, enter "why choose a CPA for accounting assistance" in your favorite search engine and review the results.

Experience

In most cases, it is essential that a foodservice operator choose an accountant that has experience working with businesses in the restaurant industry. The reason: Foodservice operations are unique businesses in that they are in both the manufacturing industry (as they prepare menu items to be sold) and the service industry (as menu items are delivered to guests).

The best accounting professionals chosen by a foodservice operator will have experience working with operations of a similar size and revenue level. This background permits the accountant to understand the unique needs of the operator.

It is a good idea to ask potential accounting professionals for a client list that details gross revenue and number of employees. This is especially important for foodservice operators who anticipate significant growth in the future. It will always be best to choose an accounting professional with the ability to continue to assist as the needs of a specific operator evolve and increase in complexity.

Assignments

Qualified accountants could be hired to handle every aspect of a foodservice operator's bookkeeping (see Chapter 1) and accounting needs. However, most accountants who are not employees charge for their services by the hour, so, to save money, some simple bookkeeping (and some accounting) tasks may be best handled in-house.

Initially, generated data about an operation's revenue and expenses should be maintained on-site where an operation's managers can regularly review it. More complex tasks such as bank account reconciliation, filling out tax return forms, payroll, and capital depreciation calculations may be delegated to the accountant. As they search for accounting assistance, foodservice operators must have a clear idea about the finance-related assignments to be completed by on-site staff and those addressed by an accounting professional.

Software Utilized

There are numerous software programs developed to assist professional accountants in their work. As a result, accountants often have their own preferred accounting software. This could result in a problem if a foodservice operator's chosen back office accounting system (see Chapter 11) cannot easily share data with the accountant's preferred system. Although it may be possible to export and import data in a suitable format, doing so can be time-consuming and easily lead to errors. In the best-case scenario, the chosen accountant's preferred software will easily interface with that of the operator.

Fees

There is no single, universal method used to establish an accountant's fees. Some accounting professionals charge by the hour and others may charge a monthly fee. Generally, the fees will be determined, at least to some degree, by the volume of work undertaken by the accountant.

When seeking professional accounting assistance, foodservice operators should require a written quotation of charges. This quotation can be compared to that of other accountants to ensure the chosen accountant's fees are reasonable and fair to both parties.

To help select an appropriate accounting professional, foodservice operators can employ several selection strategies:

1) Talk to others
2) Search online
3) Interview several candidates

1) **Talk to Others:** In many cases, a foodservice operator can utilize networks of business advisors available to provide input to decisions such as selecting the right accountant. Local Chambers of Commerce may help and might direct an operator to chamber members who voluntarily offer their advisory services for free.

   Operators who are members of their local and/or state restaurant and other related professional associations can ask their fellow members about the worth of local accounting services and if they are satisfied with their relationships. Choosing a professional accountant is most often a matter of personal preference. However, the more information operators can gather through discussions with other business professionals, the better their own decision making will likely be.

2) **Search Online:** The Internet can be a valuable tool for foodservice operators seeking professional accounting assistance. In many cases, accountants advertising their services have a fully developed website or a LinkedIn profile. Note: LinkedIn is an employment-oriented online service that operates via websites and mobile apps. The platform is primarily used for professional networking and career development; the site allows job seekers to post information about themselves, and employers can post information about available positions.

Using LinkedIn, a foodservice operator searching for an accountant can discover:

- Who are the people the accountant connects with? Do they have a strong network of other professionals?
- How do they talk about themselves? Is their LinkedIn profile professionally prepared and displayed?
- Has the accountant received any positive reviews from their current clients? What do those recommendations say?
- What is the prospective accountant's experience? How long have each been in business, and what are their areas of specialization?
- What are their qualifications? Is he/she a CPA, a bookkeeper, a financial advisor, or something else?

3) **Interview Several Candidates:** A foodservice operator's relationship with their chosen accounting professional is a key factor in the operator's success. Before making a final selection, the foodservice operator should normally interview several candidates. The specific questions to ask a candidate will vary based on the foodservice operation's needs, but in all cases a foodservice operator should ask:

1) What experience do you have with foodservice operations?

   An acceptable candidate should have relevant experience in the foodservice industry. In most cases, it would not be wise to hire a professional accountant with no foodservice industry experience.

2) How do you stay up-to-date on current accounting laws and regulations?

   Given the rate of change in legislation affecting their businesses, foodservice operators must hire an accounting professional whose priority is to stay informed. Top candidates should be able to show they monitor the latest developments, whether it is through subscriptions to industry publications, memberships in professional organizations, and/or attendance at accounting conferences and webinars.

3) Who will actually do the work?

   A senior CPA with a staff of junior accountants may be directly overseeing an operator's finances, or the CPA may do the work themselves. It is essential that operators know exactly who will be responsible for managing their financial data.

4) How accessible will you be?

   It is important to establish how easy it will be to contact the accountant. Will appointments need to be made, or will they respond to quick emails or texts? It is also important to know whether the accountant charges additional fees for routine questions and advice.

A professional accountant can be an asset for a foodservice operator, and it is very important that operators take the time to carefully choose their professional accounting partner.

Technology at Work

Some owners of foodservice operations prefer to do all accounting work in-house. Those that do so are aided in this process by using accounting software packages readily available and, in many cases, designed specifically for foodservice operations.

Accounting software is used by businesses to track income and expenses, and foodservice operators use accounting software to do many of the same things. However, software is also available to help track inventory, create invoices, calculate sales taxes, price recipes and menus, and integrate with point-of-sales (POS) systems.

In addition to preparing income statements, balance sheets, and statements of cash flow (SCF), high-quality software can also help owners and managers know details about their revenues and expenditures to identify potential savings. The software can also keep track of inventory and profits and properly calculate tip withholding amounts and sales taxes to avoid fines.

To learn more about the features included in these essential accounting systems, enter "financial accounting software packages for restaurants" in your favorite search engine and review the results.

Monitoring Evolving Accounting Issues

The foodservice industry is evolving in many ways including dining preferences, menu item popularity, and even how guests pay for their purchases. Important accounting issues also continue to evolve, and it is important for foodservice operators to continually monitor important trends and changes in accounting-related business issues.

In some cases, recommendations of how to address these emerging issues will be made by professionals in the accounting industry or established by state or federal law. In other cases, foodservice operators will be able to address these issues in the ways thought to be most helpful for their businesses.

While the following list is illustrative rather than exhaustive, it addresses some of today's accounting-related issues of growing importance to operators.

1) **Accountants on retainer:** Historically, foodservice operators have had the choice of doing their own accounting work or hiring a qualified accounting firm to do it for them. Today, with the increased numbers of easy-to-use and high-quality accounting software programs available to purchase or lease, operators can consider a third approach: pay an accountant a **retainer**.

Key Term

Retainer (professional): A fee paid in advance to secure the future services of a professional individual or firm.

A retainer is an advance payment made by a foodservice operator to a professional accountant or accounting firm. A retainer can be considered as a down payment on future requested accounting services. The actual amount paid as a retainer will vary in part based on the estimated amount of time an operator is likely to use the accountant's services.

Experienced foodservice operators know that some accounting tasks are relatively simple, while others are more complex. Unless operators have an extensive knowledge of accounting and tax-related requirements, there may be occasions when they need the assistance of a highly trained accounting professional.

For example, the fines and penalties that can occur with an incorrect tax filing or missing a tax deadline can be significant. When an accountant is on retainer, they can work with an operator to ensure the operator's business meets all tax obligations in a timely manner. In addition, the accountant may be able to identify areas of tax savings that might be unknown to the typical operator.

Other examples of potential need for a retainer agreement with an accountant include selecting the "best" depreciation method for new equipment, reviewing drafts of potentially completed income statements, and design of new tipping systems in line with changing federal and/or state regulations.

Having an accountant on retainer can also help improve the chances of obtaining a business loan. Lenders and investors carefully scrutinize the financial documents prepared by a foodservice operator seeking a loan. Having a qualified accountant on retainer helps show potential lenders an operator is serious about the quality of the financial information being provided to potential sources of external funding.

2) **Properly accounting for third-party delivery fees:** The use of third-party delivery services by foodservice operators is not new, but the COVID-19 pandemic and resulting changes in consumer behavior have made these fees larger and more significant to operators than ever before.

It is likely that the relationship between foodservice operators and their chosen third-party delivery services will continue to evolve. From an accounting perspective, foodservice operators can consider various ways to account for these fees including:

a) Reducing third-party fees from sales. For example, a $50 guest check with $10 of third-party delivery fees could be posted as $40 sales rather than $50 since $40 is the amount the operator receives from the sale.

b) Recording the full amount of a guest's purchase as sales and posting the cost of third-party delivery fees as a cost of sales expense. For example, a $50 guest check including $10 of third-party delivery fees could be posted as $50 revenue with a $10 expense charge to a designated cost of sales expense category.

c) Recording the full amount as sales and posting the cost of third-party delivery fees as a marketing expense. For example, a $50 guest check with $10 of third-party delivery fees might be posted as $50 revenue with a $10 expense charge to marketing.

There are accounting- and sales tax-related issues associated with each of the above approaches, but it is important to remember that the purpose of accounting is to reflect the true revenue and costs of operating a business. Third-party delivery fees are an excellent example of how accounting methods must evolve to address changes in methods used by business operations.

3) **Charging extra for guests who pay with credit cards:** In March 2017, the U.S. Supreme Court ruled that, in nearly all states, businesses including foodservice operators could charge customers paying bills with credit cards more than they charge guests who pay their bills with cash or debit cards.[1]

The ruling applies only to credit (not debit) cards and, while state laws vary somewhat, in most cases operators who implement this action must:

- Give guests notice before their purchases are made. Note: This can be in the form of a clear written notice on the menu, or a verbal statement provided by service staff.
- Disclose the surcharge on the guest's receipt.
- Post a general notice of this practice at the entrance door and on the operation's website.

Increasingly, foodservice operators (especially smaller ones) are adopting this practice. While several accounting-related recording issues are associated with this decision, one aspect that is often overlooked involves guest perception.

Some guests likely have no objection to paying a small fee for the use of their credit cards. In other cases, however, the reaction may be negative. Foodservice accountants must provide employees who deal directly with guests required tools to address any customer complaints that arise from implementing this practice.

---

[1] https://fortune.com/2017/03/29/credit-card-charges-supreme-court-freedom-speech/, retrieved January 31, 2023.

4) **Properly accounting for costs related to employee recruitment, selection, and retention:** Historically, many foodservice operators categorized their expenses for employee recruitment, selection, and training as costs within their Human Resources (HR) departments. Some operators classified these expenses as a marketing expense because the search for new employees also involved marketing the operation and its desirability as a good place to work.

Finally, many smaller operators recorded employee hiring and retention-related costs as a general and administrative (G&A) expense because employees are necessary to operate the business.

The authors argue that foodservice operators should consider a unique category within labor costs to record these expenses. As labor markets tighten and staff availability is reduced, higher proportions of revenue will need to be devoted to recruiting and retaining a productive workforce. As a result, thoughtful operators could consider categorizing these expenses as a part of labor costs, and this can be done within the Uniform System of Accounts for Restaurants (USAR) recommendations.

5) **Accounting for automatic gratuity/service charges:** For numerous reasons, foodservice operators are including automatic gratuities on their guests' checks. Experienced managerial accountants in the foodservice industry know that automatic gratuities charged by an operation are significantly different from tips. Increasingly, these charges are determined by the POS system and added automatically to guests' checks based on percentages or amounts established by the foodservice operator.

Since this practice takes away the customer's ability to determine the gratuity amount, the Internal Revenue Service (IRS) considers these to be **service charges** rather than tips. Note: The IRS considers service charges to be revenue, and they are considered "sales" (and generally subject to state sales taxes). If all or a portion of the service charges are dispersed to staff, however, the disbursement amounts must be considered wages and are subject to withholding tax (while tips generally are not).

Key Term

Service charge: The amount a guest must pay under the terms and conditions of purchasing food and drinks in a foodservice operation. Service charges belong to the operation and not to the operation's employees.

Recent court rulings have held that "whether the restaurant included the service charges in its gross receipts on its tax returns or paid sales taxes on the amounts was "irrelevant" to determining whether the service charges were wages or tips."[2]

---

2 https://ogletree.com/insights/eleventh-circuit-service-charges-are-wages-not-tips-under-flsa/, retrieved January 31, 2023.

6) **Movement to cloud-based accounting** Cloud-based accounting allows a foodservice operation to minimize the amount of accounting-related hardware and software it must purchase and maintain on-site. Moving data storage to the cloud will be increasingly cost-effective. However, it is also true that data breaches are an increasingly greater risk.

Targets of data breaches include the stealing of an operation's financial and personal data including guests' credit card information. As a result, cyber security will take on greater importance as the movement to cloud-based accounting increases.

Foodservice operators interested in properly documenting their financial activities must continually monitor changes in the business environment that directly affect their accounting procedures. Issues such as those explored above (and many new ones!) will likely continue to shape how foodservice operators prepare financial documents that reflect operating results.

What will not likely change, however, is the interest of operators in accurately documenting their business success, and in using financial data and their own insight to help make decisions that will assist them in achieving even greater success in the future.

What Would You Do? 12.2

"Well, they both seem to be good, but they also seem to be very different," said Ericka.

Ericka and Ricky, her partner in the "Bake It To The Limit" bakery were talking about the two accounting firms they had interviewed to help with their accounting needs.

Ricky was responsible for marketing the bakery, and Erica was initially responsible for managing production and keeping the bakeshop's books. Since Ricky has been successful, the bakeshop has become very popular. The problem: Ericka has needed to spend much more time with production tasks, and she now has less time to do "office work." For that reason, they were seeking assistance in managing their accounting data.

"Larson's Accounting Services will charge us more," said Ricky. "They are a big firm, and they have a ton of CPAs on staff. I really feel like we'd be safe choosing them."

"Yes," replied Ericka, "that's true. But I think Tony Guild's Accounting might be better for us. I know the company is only Tony and his one office assistant, and it's really small. However, Tony will charge us less, and he deals exclusively with businesses in the foodservice industry."

Assume you were Ericka and Ricky. How important do you believe it would be to have your accounting work performed by a firm that employs CPAs? How important do you believe it would be to have your work done by a firm with extensive experience in the foodservice industry? Explain your answers.

Key Terms

Capital expenditure

Revenue expenditure

Capital lease

Straight-line depreciation

Double declining balance depreciation (DDB)

Accelerated method of depreciation

Financial Accounting Standards Board (FASB)

Beneficiary

Cash surrender value

Cloud accounting software

Certified Public Accountant (CPA)

Retainer (professional)

Service charge

Operator's 10-Point Tactics for Success Checklist

Evaluate your need for, and the current status of, each of the following operational tactics. For those tactics you think are important, but not yet in place, develop an action plan for its implementation including who will be responsible for the tactic's completion and the target date by which it should be completed.

Tactic	Don't Agree (Not Done)	Agree (Done)	Agree (Not Done)	If Not Done	
				Who Is Responsible?	Target Completion Date
1) Operator understands that expenditures for fixed assets are recorded as capital expenditures rather than revenue expenditures.	⎯	⎯	⎯		
2) Operator lists the recorded value of newly purchased fixed assets to include the asset's original cost plus taxes, shipping, installation, and any other applicable costs.	⎯	⎯	⎯		
3) Operator has, if needed, received professional advice about the use of straight-line and/or double declining balance methods for computing depreciation for applicable fixed assets.	⎯	⎯	⎯		

Tactic	Don't Agree (Not Done)	Agree (Done)	Agree (Not Done)	If Not Done	
				Who Is Responsible?	Target Completion Date
4) Operator's net book values of all property and equipment are kept current using their selected depreciation method(s).	____	____	____		
5) Where it is practical, operator capitalizes the initial cost of uniforms, linen, china, glass, silver, and utensils.	____	____	____		
6) Replacement costs for uniforms, linen, china, glass, silver, and utensils are considered an expense in the month they are placed into service.	____	____	____		
7) Operator recognizes the sale or exchange of a fixed asset results in an adjustment in the appropriate accumulated depreciation and/or fixed asset account.	____	____	____		
8) Operator recognizes that all "Other Assets" owned by the operation must be clearly indicated on the operation's balance sheet.	____	____	____		
9) Operator has carefully evaluated the need for obtaining professional assistance in the preparation of accounting records.	____	____	____		
10) Operator has implemented a well-designed procedure for selecting professional assistance in the preparation of accounting records.	____	____	____		

Glossary

28-day accounting period: An accounting period that is four weeks (28 days) in length instead of a calendar month that has between 28-31 days. There are 13 four-week periods instead of 12 monthly periods when using this system.

Accelerated method of depreciation: A method of depreciation that results in higher depreciation charges in the first year; charges gradually decline over the life of a fixed asset.

Account balance: The total amount of money available in a financial account after all the debits and credits have been calculated.

Account: An accounting device that shows increases and decreases in a single asset, liability, or owners' equity item.

Accountant: An individual skilled in the recording and reporting of financial transactions.

Accounting period: The amount of time included on a financial summary or report, and which should be clearly identified on the financial document.

Accounting: The system of recording and summarizing financial transactions and analyzing, verifying, and reporting the results.

Accounts payable (AP): Money owed by the foodservice operation to suppliers and lenders that has not yet been paid. Sometimes referred to as "AP."

Accounts receivable (AR): Money owed to a foodservice operation, generally from guests, which has not yet been received. Sometimes referred to as "AR."

Accrual accounting system: An accounting system that matches expenses incurred with revenues generated. This is done with the use of accounts receivable, accounts payable, and other similar accounts.

Accumulated depreciation: The sum of all recorded depreciation on an asset to a specific date.

Activity-based costing: A method of assigning overhead and indirect costs such as salaries and utilities to specific products and services.

Aggregate statement: A financial statement in which data from many individual financial accounts are summarized in one document.

Allowance for doubtful accounts: A reserve held for probable losses when all accounts receivable are not collected.

Amortization: The practice of spreading an intangible asset's cost over that asset's useful life.

Appreciation: An increase in the value of a business asset over time. Unlike depreciation, which lowers an asset's value over its useful life, appreciation is the rate at which an asset grows in value.

Assets: Something of value owned by a foodservice operation. Examples include cash, product inventories, equipment, land, and building(s).

Auditing: The accounting specialty that involves studying a foodservice operation's internal controls and analyzing the basic accounting system to ensure that all financial information is properly recorded and reported.

Auditor: An accounting professional who specializes in auditing.

Back office accounting system: The accounting software used to maintain a business's accounting records that are not contained in its point of sale (POS) system. This typically includes items such as payroll records, accounts payable and receivable, taxes due and payable, and net profit and loss summaries, as well as all balance sheet entries.

Balance sheet: A report that documents the assets, liabilities, and net worth (owners' equity) of a foodservice business at a single point in time. Also commonly called the Statement of Financial Position.

Bank reconciliation: A process performed to ensure that a company's financial records are correct. This is done by comparing the company's internally recorded amounts with the amounts shown on its bank statement. Any differences must be justified. When there are no unexplained differences, accountants can affirm that the bank statement has been reconciled.

Bankruptcy: A legal proceeding initiated when a person or business is unable to repay outstanding debts or obligations.

Basic accounting equation: A formula which shows that a company's total assets are equal to the sum of its liabilities and its shareholders' equity.

Benchmark: Something serving as a standard by which others may be measured or judged.

Beneficiary: A person or entity who is the recipient of funds or other property under a will, trust, or insurance policy.

Bonding (employee): Protection from loss resulting from a crime event involving employee dishonesty and theft such as loss of money, securities, and other property.

Bookkeeping: The process of recording a foodservice operation's financial transactions into organized accounts on a daily basis.

Break-even point: The sales point at which total cost and total revenue are equal: there is no loss or gain for a business.

Bundling: A pricing strategy that combines multiple menu items into a grouping which is then sold at a price lower than that of the bundled items purchased separately.

Business dining: The non-commercial segment of the foodservice industry that serves the dining needs of large businesses and corporations.

Calendar year: A 365-day time period that begins on January 1st and ends on December 31st. of the same year.

Capital expenditures: Expenditures made by a foodservice operation in which benefits are expected to be received for a period greater than one year.

Capital lease: A lease agreement of long duration, generally noncancelable, in which the lessee (foodservice operation) assumes responsibility for property taxes, insurance, and maintenance of the leased property.

Capital stock: The total number of shares of stock a company is legally authorized to issue.

Cash equivalent: A short-term, temporary investment such as treasury bills, certificates of deposit, or commercial paper that can be quickly and easily converted to cash.

Cash flows: The net amount of cash and cash equivalents being transferred in and out of a business. Cash received represents inflows, and money that is spent represents outflows.

Cash management: The process used to effectively control a business's cash balances (currency on hand and demand deposits in banks), cash flow (cash receipts and disbursements), and short-term investments in securities.

Cash on hand: The amount of money a business has available to spend on the last day of an accounting period.

Cash receipts journal: An accounting record used to summarize transactions related to cash receipts generated by a foodservice operation.

Cash surrender value: Money an insurance company pays to a policyholder if the policy is voluntarily terminated before maturity, or an insured event occurs.

Cash: Any medium of exchange a bank will accept at face value.

Certified public accountant (CPA): An accounting professional who has earned a CPA license through a combination of education, experience, and examination.

Chief Financial Officer (CFO): A senior executive responsible for managing the financial actions of a company.

Classification (expense): Determining where to place an expense on an income statement.

Closed out (account): The accounting steps used to transfer amounts from temporary accounts to permanent accounts.

Cloud accounting software: Software like traditional, on-premises, or self-installed accounting software, except the accounting software is hosted on remote servers. Data is sent into "the cloud," where it is processed and returned to the user.

Common stock: A security that represents ownership in a corporation. Holders of common stock elect the company's board of directors and can vote on corporate policies.

Condensed income statement: An income statement that reduces much of the normal income statement detail to just a few lines. Typically, all revenue line items are aggregated into a single line item, while the cost of sales appears as one line item. Labor and all other operating expenses appear in separate lines.

Consolidated income statement: A single report that combines revenue, expense, and profit information for two or more affiliated business operations.

Consumer rationality: The tendency to make buying decisions based on the belief that the decisions are of personal benefit.

Contra asset (account): An account type used to decrease the balance in an asset account. The account is not classified as an asset since it does not represent a long-term value. It is not classified as a liability since it does not constitute a future obligation.

Contribution margin income statement: An income statement in which all variable expenses are deducted from sales to arrive at a contribution margin. Then, all fixed expenses are subtracted to arrive at the net profit or net loss for the accounting period.

Contribution margin: The dollar amount remaining after subtracting variable costs from total sales that *contributes* to covering fixed costs and providing for a profit.

Controllable costs: Those costs over which on-site foodservice managers have primary control.

Controllable Income: The amount of money remaining after an operation's controllable expenses have been subtracted from its total sales.

Controllable labor expenses: Those labor expenses under the direct influence of management.

Cost accounting: The accounting specialty that is involved with classifying, recording, and reporting a business's expenses or costs.

Cost center: A part of a business organization with assignable expenses that generates little or no revenue.

Cost of sales: The total cost of the products used to make the menu items sold by a foodservice operation.

Count (product): A number used to designate product size or quantity when purchasing food items.

CPA: Short for Certified Public Accountant. A CPA is a professional designation given to qualified accountants who have passed a rigorous accounting exam (the Uniform CPA Exam).

Credit (entry): Any entry made on the right side of a "T" account.

Credit memo: An adjustment to a vendor's delivery invoice that reconciles any differences between the delivery invoice, the products ordered (PO), and products received.

Creditor: A person or company to whom money is owed.

Current fair value: The measuring of a business's liabilities and assets at their current market value; the amount that an asset could be sold for (or that a liability could be settled for) that is fair to both buyer and seller.

Current liability: The total of all obligations due to be paid within 12 months of the balance sheet date.

Current ratio: The liquidity ratio that compares current assets to current liabilities and indicates an operation's ability to pay its current obligations in a reasonable time.

Data privacy: The aspect of data management that addresses who has ownership over data, who can determine its legitimate use, and the regulations related to utilizing this data.

Data security: The steps taken, and the safeguards implemented to prevent unauthorized access to an organization's data.

Debit (entry): Any entry made on the left side of a "T" account.

Debt to equity ratio: A ratio that compares a business's equity to its total liabilities.

Delivery invoice: A detailed listing of products and their costs delivered by a vendor.

Depreciation: The allocation of the cost of equipment and other depreciable assets based on the projected length of their useful life.

Direct cost: A cost that most often increases as activity or the volume within a revenue center increases.

Dividends: A reward paid to the shareholders of a company for their investment in the company's equity. Dividends usually originate from a company's net profits.

Double declining balance depreciation (DDB): An accelerated method of depreciation that ignores the salvage value of a fixed asset and yields a greater amount of depreciation in the early years of the asset's life.

Double-entry accounting: A bookkeeping and accounting system that requires the recording of each business transaction in at least two accounts. While its use is not mandatory for private businesses, all public companies (those that issue stock) are required to use double-entry accounting to meet GAAP requirements. Also commonly called double-entry bookkeeping.

EBITDA: Short for: "Earnings before interest, taxes, depreciation, and amortization." EBITDA is used to track and compare the underlying profitability of a business regardless of its depreciation assumptions or financing choices.

Embezzlement: A crime in which a person or entity intentionally misappropriates assets that have been entrusted to them.

Employee benefits: Any form of rewards or compensation provided to employees in addition to their base salaries and wages. Benefits offered to employees may be legally mandatory or voluntary.

Empowerment (employee): An operating philosophy that emphasizes the importance of allowing employees to make independent guest-related decisions and to act on them.

Equities: The claims against a foodservice operation's assets by those who provided the assets.

Ethics: A system or code of moral rules, principles, or values.

Exempt employee: Employees exempted from the provisions of the Fair Labor Standards Act. These workers typically must be paid a salary above a certain level and work in an administrative or professional position. The U.S. Department of Labor (DOL) regularly publishes a duties test that can help employers determine who meets this exemption.

Expense: A decrease in a resource, such as food inventory, which occurs when a foodservice operation sells a product or service or incurs a business cost.

Fast casual (restaurant): A sit-down foodservice operation with no wait staff or table service. Customers typically order off a menu board and seat themselves or take the purchased food elsewhere.

Favorable variance: Better than expected performance when actual results are compared to budgeted results.

Financial accounting: The process of developing and using accounting information to make business decisions. It involves organizing and presenting financial information in financial statements and summaries, and its major focus is on the past.

Financial Accounting Standards Board (FASB): An independent, private-sector, not-for-profit organization that establishes financial accounting and reporting standards for public and private companies and not-for-profit organizations that follow Generally Accepted Accounting Principles (GAAP).

Financial returns: The money made or lost on an investment over a specific time.

First-in First-out (FIFO) inventory system: An inventory management system in which products already in storage are used before more recently delivered products.

Fiscal year: A time period that can begin on any date and then concludes 365 consecutive days after it begins.

Fixed asset: An asset such as land, building, furniture and equipment which is purchased for long-term use and is not likely to be converted quickly into cash.

Fixed cost: An expense that stays constant despite increases or decreases in sales volume.

Fixed payroll: The amount an operation pays for its salaried workers.

Food cost percentage: A ratio calculated by determining the food cost for a menu item and dividing that cost by the item's selling price.

Franchised operation: A method of distributing products or services involving a franchisor, who establishes the brand's trademark or trade name and a business system, and a franchisee, who pays a royalty and often an initial fee for the right to operate under the franchisor's name and system.

Fraud: Deceitful conduct to manipulate someone into giving up something of value by (a) presenting as true something known by the fraudulent party to be false or (b) by concealing a fact from someone that may have saved him or her from being cheated.

Free cash flow: The amount of cash a business generates from its operating activities *minus* the amount of cash it must spend on its investment activities and capital expenditures.

Full-service (restaurant): A foodservice operation at which servers deliver food and drink offered from a printed menu to guests seated at tables or booths.

Furniture, fixtures, and equipment (FF&E): Movable furniture, fixtures, or other equipment that have no permanent connection to a building's structure.

General journal: Also referred to simply as the "journal," this accounting document is involved in the first phase of accounting because all transactions are initially recorded in it, originally and in chronological order. General journals are also often referred to as "the journal," an "individual journal," or "the book of original entry."

General ledger: The main or primary accounting record of a business.

Generally Accepted Accounting Principles (GAAP): Standards that have evolved in the accounting profession to ensure uniformity in the procedures and techniques used to prepare financial statements.

Going concern: A business entity that generates enough cash to stay in business.

Goodwill (Other Asset): The sum of all special advantages relating to the business including its reputation, well-trained staff, highly motivated management, and favorable location.

Guest check average: The average (mean) amount of money spent per guest (or table) during a specific accounting period. Also referred to as "check average" or "ticket average."

Historical cost: A measure of value used in accounting in which the value of an asset is recorded at its original cost when acquired by the company. The historical cost method is used for fixed assets under generally accepted accounting principles (GAAP).

Historical data: Information about an operation's past financial performance including revenue generated, expenses incurred, and profits (or losses) realized.

Income before income taxes: The amount of money remaining after an operation's interest expense is subtracted from the amount of its restaurant operating income. Also, a business's profit before paying any income taxes due on the profits.

Income statement: Formally known as "The Statement of Income and Expense," a report summarizing a foodservice operation's profitability including details regarding revenue, expenses, and profit (or loss) incurred during a specific accounting period. Also commonly called the Profit and Loss (P&L) statement.

Incremental cost: The total cost incurred due to an additional unit of product being produced or a guest being served.

Indirect cost: A cost that is cannot be readily assigned to a specific revenue center in a foodservice operation. Also referred to as an overhead cost.

Interest expense: The cost of borrowing money.

Issuing system: The procedures in place to control the removal of products from storage.

Job description: A statement that outlines the specifics of a particular job or position within a company. It provides details about the responsibilities and conditions of the job.

Job specification: A listing of the personal characteristics and skills needed to perform the tasks contained in a job description.

Joint cost: A cost that benefits more than one product or profit center. Also referred to as a "shared" cost.

Journal: A book for original entry of financial information into the accounting system.

Journal entry: The first step in the accounting cycle, and it is a record of the financial transaction in the accounting books of a business. A properly documented journal entry consists of the correct date, amounts entered, description of the transaction, and a unique reference number.

Labor cost percentage: A ratio of overall labor costs incurred relative to total revenue generated.

Labor dollars per guest served: The dollar amount of labor expense incurred to serve each of an operation's guests.

Leasehold improvement: Anything that benefits one specific tenant in a rental arrangement. Examples include painting, adding new walls, putting up display shelves, changing flooring and lighting, and adding walls and partitions.

Liability: Obligations (money owed) to outside entities. Examples include amounts owed to suppliers for delivered products, to lenders for long-term debt such as a mortgage, and to employees (payroll) that has been earned by, but not yet paid to, an operation's staff members.

Lien: A legal right acquired on one's property by a creditor. A lien generally stays in effect until the underlying obligation to the creditor is satisfied. If the underlying obligation is not satisfied, the creditor may be able to take possession of the property involved.

Liquidity: A measure of a foodservice operation's ability to convert assets into cash.

Liquidity ratios: A category of ratios that indicate a business's ability to pay current obligations.

Long-term debt: The category of liabilities related to debt that will not be paid within 12 months of the balance sheet date.

Malware: Software that is specifically designed to disrupt, damage, or gain unauthorized access to a computer system.

Managerial accounting: The accounting specialty that uses historical and estimated financial information to help foodservice operators plan the future.

Menu engineering: A system used to evaluate menu pricing and design by categorizing each menu item into one of four categories based on its profitability and popularity.

Merchant services provider: A company that enables foodservice operators (merchants) to accept credit and debit card payments and other alternative payment methods. Also sometimes referred to as a "Payment services provider."

Minimum operating cost: The least dollar amount of cost of sales plus variable costs incurred by a foodservice operation during a defined time period.

Minimum sales point (MSP): The dollar amount of sales needed for a business to profitably remain open during a defined time period.

Mixed cost: A cost composed of a mixture of fixed and variable components. Costs are fixed for a set level of sales volume or activity and then become variable after this level is exceeded. Also referred to as a semi-fixed or semi-variable cost.

Net book value: The cost of a fixed asset minus its accumulated depreciation.

Net changes in cash: The total increase or decrease of cash and cash equivalent balances within a specified accounting period factoring in the net changes in cash for operating, investing, and financing activities.

Net income: A number calculated as revenue minus operating expenses including depreciation, interest, and taxes, among others. It is a useful to assess how much revenue exceeds the expenses of operating a business in a defined time period.

Non-cash expense: Expenses recorded on the income statement that do not involve an actual cash transaction. Examples include depreciation and amortization which are expenses where an income statement charge reduces operating income without a cash payment.

Noncash transaction: Investing and financing-related transactions that do not involve the use of cash or a cash equivalent.

Non-commercial foodservice: A foodservice operation created primarily to support an organization's larger mission such as education, healthcare, or business/industry.

Non-controllable expenses: Costs which, in the short run, cannot be avoided or altered by management decisions. Examples include lease payments and depreciation.

Non-controllable labor expenses: Those labor expenses not typically under the direct influence of management.

Notes to the Financial Statements: A supplemental narrative statement managerial accountants attach to a financial report to provide useful and critical information that would otherwise not be available.

Occupancy costs: Costs related to occupying a space including rent, real estate taxes, personal property taxes, and insurance on a building and its contents.

Open check: A guest check that initially authorizes product issues from the kitchen or bar, but has not been collected for (closed) and, therefore, has not been added to the operation's sales total.

Operating budget: An estimate of the income, expenses, and profit for a business over a defined accounting period. Also referred to as a "Financial plan."

Opportunity cost: The potential benefits that an investor misses out on when selecting one alternative course of action over another.

Other Controllable Expenses: Expenses that a foodservice operator can influence with increases or decreases based on business decisions. Examples include marketing costs and utility costs.

Over (cash bank): A cashier bank that, based on actual sales, contains more than the amount it should contain.

Owners' equity: The assets of an operation minus its liabilities. Also, the net financial interest of an operation's owner(s).

Par value (stock): The value of a single common share of stock as set by a corporation's charter. It is not typically related to the actual value of the shares. In fact, it is most often lower.

Payroll: The term commonly used to indicate the amount spent for labor in a foodservice operation. Used for example in: "*Last month our total payroll was $28,000.*"

Permanent owners' equity account: An owners' equity account in which the balance at the end of an accounting period becomes the beginning balance for the next accounting period. Sometimes referred to as "real" accounts.

Petty cash: Money kept on hand to purchase minor items when the use of a check or credit card is not practical.

Phishing: The fraudulent practice of sending emails supposedly from reputable companies to induce individuals to reveal personal information including passwords and credit card numbers.

Point-of-sale (POS) system: An electronic system that records foodservice customer purchases and payments, as well as other operational data.

Popularity Index: The proportion of guests choosing a specific menu item from a list of alternative menu items. The formula for a popularity index is:

$$\frac{\text{Total Number of a Specific Menu Item Sold}}{\text{Total Number of All Menu Items Sold}} = \text{Popularity Index}$$

Portion cost: The product cost required to produce one serving of a menu item.

Posting: Moving a transaction entry from a journal to a general ledger.

Potentially hazardous food (PHF): Foods that must be kept at a particular temperature to minimize the growth of food poisoning bacteria or to stop the formation of harmful toxins.

Price (Noun): A measure of the value given up (exchanged) by a buyer and a seller in a business transaction.

Price (Verb): To establish the value to be given up (exchanged) by a buyer and a seller in a business transaction.

Prime cost: An operation's cost of sales plus its total labor costs.

Principle (loan): The amount a borrower agrees to pay a lender when a loan becomes due, not including interest.

Product specification (spec): A detailed description of a recipe ingredient or complete menu item to be served.

Productivity (worker): The amount of work performed by an employee within a fixed time period.

Profit and Loss (P&L) statement: Formally known as "The Statement of Income and Expense," this is a report summarizing a foodservice operation's profitability including details about revenue, expenses, and profit (or loss) incurred during a specific accounting period. Also commonly referred to as the "Income Statement."

Profitability ratios: A group of ratios that assess a business's ability to generate earnings relative to its revenue, operating costs, balance sheet assets, or shareholders' equity.

Profit center: A part of a business organization with assignable revenues and expenses. Also referred to as a "revenue center."

Profit margin: The amount by which revenue in a foodservice operation exceeds its operating and other costs.

Purchase order (PO): A detailed listing of products a buyer is requesting from a vendor.

Quick service (restaurant): Foodservice operations which typically have limited menus and often include a counter at which customers can order and pick up their food. Most quick service restaurants also have one or more drive-through lanes which allow customers to purchase menu items without leaving their vehicles.

Ratio analysis: The comparison of related information, most of which is found on financial statements.

Ratio: An expression of the relationship between two numbers; computed by dividing one number by the other number.

Restaurant operating income: All an operation's revenue minus its controllable and non-controllable expenses.

Restaurant row: A street or region well-known for having multiple foodservice operations within close proximity.

Retained earnings: The cumulative net (retained) earnings or profits of a company after accounting for any dividends that have been paid out. Retained earnings decrease when a company loses money or pays out dividends, and they increase when new profits are generated.

Retainer (professional): A fee paid in advance to secure the future services of a professional individual or firm.

Return on assets (ROA): A financial ratio indicating the profitability of a company in relation to total assets.

Return on equity (ROE): The measure of a company's net income divided by its shareholders' equity. ROE is a gauge of a corporation's profitability and how efficiently it generates those profits. The higher the ROE, the better a company is at converting its equity financing into profits.

Return on investment (ROI): A measure of the ability of an investment to generate income.

Revenue: An increase in a resource such as cash which occurs when a product or service is sold by a business. Also commonly referred to as "sales," or "income."

Revenue expenditures: Expenditures made by a foodservice operation in which benefits are expected to be received within one year.

Salaried employee: An employee who regularly receives a predetermined amount of compensation each pay period on a weekly, or less frequent, basis. The predetermined amount paid is not reduced because of variations in the quality or quantity of the employee's work.

Sales forecast: An estimate of the number of guests to be served and the menu items they will purchase within a specific time period.

Sales mix: The series of individual guest purchasing decisions that result in a specific overall food or beverage cost percentage.

Sales per labor hour: The dollar value of sales generated for each labor hour used.

Salvage value: The estimated resale value of an asset at the end of its useful life. Salvage value is subtracted from the cost of a fixed asset to determine the amount of the asset cost to be depreciated so salvage value is a component of the depreciation calculation.

Service charge: The amount a guest must pay under the terms and conditions of purchasing food and drinks in a foodservice operation. Service charges belong to the operation and not to the operation's employees.

Short (cash bank): A cashier bank that, based on actual sales, contains less than the amount it should contain.

Significant variance: A variance that requires immediate management attention.

Smishing (attack): The fraudulent practice of sending text messages supposedly from reputable companies to induce individuals to reveal personal information such as passwords or credit card numbers.

Solvency: The ability of a business to meet its long-term financial obligations as they become due.

Solvency ratio: A ratio that measures total assets against total liabilities.

Sources and uses of funds: Inflows and outflows of money affecting a business's cash position.

Split-shift: A working schedule comprising two or more separate periods of duty in the same day.

Stakeholder: A person or organization with a vested interest in a foodservice operation that can either affect or be affected by the operation's financial performance.

Standard cost: The estimated expense that occurs during the production of a product or performance of a service. Also referred to as an "expected" cost, a "budgeted" cost, or a "forecasted" cost.

Standardized recipe: The instructions needed to consistently prepare a specified quantity of food or drink at an expected quality level.

Statement of Cash Flows (SCF): A report providing information about all cash inflows a company receives from its ongoing operations and external investment sources. It also includes all cash outflows paying for business activities and investments during a given accounting period.

Statement of retained earnings: A financial document that reports the changes in a business's retained earnings from last year to this year.

Step costs: Costs that increase in a nonlinear fashion as activity or volume increases.

Stockholders' equity: A claim to assets of a corporately-owned foodservice operation by the corporation's owners (stockholders).

Straight-line depreciation: A method of depreciation that distributes the expense evenly over the estimated life of a fixed asset.

Sunk cost: Money that has already been spent and cannot be recovered.

Supporting schedule: A detailed itemization of the contents of an account.

"T" account: An informal term for a set of financial records that use double-entry bookkeeping It is called a "T" account because the bookkeeping entries are laid out in a way that resembles a "T" shape. The title (name) of the account appears just above the "T."

Target market: The group of people with one or more shared characteristics that an operation has identified as most likely customers for its products and services.

Tax accounting: The accounting specialty that involves planning and preparing for required tax payments and filing tax-related information with governmental agencies.

Temporary owners' equity account: This includes income statement accounts (revenues, expenses, gains, and losses), the owner's drawing account (used to record the amounts withdrawn from a sole proprietorship by its owner), and the income summary accounts. These are temporary owner's equity accounts because, at the end of the year the balances in these accounts are transferred to the owner's permanent capital account(s).

Third-party delivery: The use of a smart phone or computer application that allows customers to browse restaurant menus, place orders, and have them delivered to the customer's location. In nearly all cases the requested orders are delivered by independent contractors retained by the company operating the third-party delivery app.

Timing (expense): Determining when to place an expense on an income statement.

Tip distribution program: A system of tip payment that allows an operation to distribute a customer's tip from an employee who actually received it to others who also provided service to the customer.

Tip pooling: A tip system that takes the tips given to individual employees in a group and shares them equally with all other members of the group. Used for example, when bartender tips given to an individual bartender are shared equally with all bartenders working the same shift.

Tip sharing: A tip system that takes the tips given to one group of employees and gives a portion of them to another group of employees: Used, for example, when server tips are shared with those who bus the server's tables.

Topline revenue: Sales or revenue shown on the top of the income statement of a business.

Total labor: The cost of the management, staff, and employee benefits expense required to operate a business.

Total sales: The sum of food sales and alcoholic beverage sales generated in a foodservice operation.

Transaction: Any business event having a monetary impact on the financial statements of a business.

Unfavorable variance: Worse than expected performance when actual results are compared to budgeted results.

Uniform system of accounts: Accounting standards used to provide uniformity and consistency in reporting financial information.

Uniform System of Accounts for Restaurants (USAR): A recommended and standardized (uniform) set of accounting procedures used for categorizing and reporting restaurant revenue and expenses.

Value: The amount paid for a product or service compared to the buyer's view of what they receive in return.

Variable cost: An expense that generally increases as sales volume increases and decreases as sales volume decreases.

Variable payroll: The amount an operation pays for those workers compensated based on the number of hours worked.

Variance (budget): The difference between an operation's actual performance and its budgeted (planned) performance.

Vertical analysis: A method of financial statement analysis in which each line item is listed as a percentage of a base figure within the statement. Also known as "common-size" analysis.

Vishing: The fraudulent practice of making phone calls or leaving voice messages purporting to be from reputable companies to induce individuals to reveal personal information such as bank details and credit card numbers.

Wagyu beef: Beef from a Japanese breed of cattle that is highly prized for its marbling and flavor. In the Japanese language, "Wa" means Japanese, and "gyu" means cow.

Walk (bill): Guest theft that occurs when a guest consumes products that were ordered and leaves without paying the bill. Also referred to as a "Skip."

Weighted contribution margin: The contribution margin provided by all menu items divided by the total number of items sold. Weighted contribution margin is calculated as:

$$\frac{\text{Total Contribution Margin of All Items Sold}}{\text{Total Number of Items Sold}} = \text{Weighted contribution margin}$$

Working capital: The capital of a business used in its day-to-day operations.

Index